Proceedings of the Conference on

Stochastic Differential Equations and Applications

Academic Press Rapid Manuscript Reproduction

Proceedings of a conference on
Stochastic Differential Equations and Applications
held in Park City, Utah, on February 17–20, 1976

Proceedings of the Conference on

Stochastic Differential Equations and Applications

Edited by

J. DAVID MASON

Department of Mathematics
University of Utah
Salt Lake City, Utah

Academic Press, Inc.

New York San Francisco London 1977

A Subsidiary of Harcourt Brace Jovanovich, Publishers

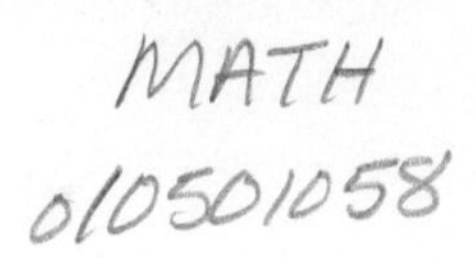

This work relates to the Department of the Navy Contract N00014-76-G-0030 issued by the Office of Naval Research under Contract Authority NR 041-499. However, the content does not necessarily reflect the position or the policy of the Department of the Navy or the Government, and no official endorsement should be inferred.

ACADEMIC PRESS, INC.
111 Fifth Avenue, New York, New York 10003

United Kingdom Edition published by
ACADEMIC PRESS, INC. (LONDON) LTD.
24/28 Oval Road, London NW1

Library of Congress Cataloging in Publication Data

Conference on Stochastic Differential Equations and Applications, Park City, Utah, 1976.
Proceedings of the Conference on Stochastic Differential Equations and Applications.

"Department of the Navy contract N00014-76-G-0030 issued by the Office of Naval Research."
1. Stochastic differential equations—Congresses.
I. Mason, Jesse David.
QA274.23.C36 1976 519.2'4 77-8503
ISBN 0-12-478050-4

PRINTED IN THE UNITED STATES OF AMERICA

Contents

List of Participants vii
Preface ix

The Deterministic Itô-Belated Integral Is Equivalent to the Lebesgue Integral 1
Richard B. Darst

Optimal Stopping and Quasi Variational Inequalities 5
Avner Friedman

The Regular Expansion in Singularly Perturbed Elliptic Equations 23
Charles J. Holland

Fluctuation Phenomena in Underwater Sound Propagation, I 45
Werner Kohler and George C. Papanicolaou

Fluctuation Phenomena in Underwater Sound Propagation, II 113
Werner Kohler and George C. Papanicolaou

Relations between Sample and Moment Stability for Linear Stochastic Differential Equations 145
F. Kozin and S. Sugimoto

Diffusion Approximations in Population Genetics—How Good Are They? 163
Benny Levikson

Some Unsolved Problems and New Problem Areas in the Field of Stochastic Differential Equations 189
James McKenna

Diffusion on the Line and Additive Functionals of Brownian Motion 211
Steven Orey

An Individual Ergodic Theorem for the Diffusion on a Manifold of Negative Curvature **231**
Mark A. Pinsky

Two Limit Theorems for Random Evolutions Having Non Ergodic Driving Processes **241**
D. Stroock

List of Participants

George Adomian	The University of Georgia
Jerry Bona	The University of Chicago
Neil Bromberg	Rochester Institute of Technology
Richard B. Darst	Colorado State University
Philip Feinsilver	The University of Utah
Neil Fenichel	University of British Columbia
Avner Friedman	Northwestern University
Richard Griego	The University of New Mexico
Alberto Grünbaum	University of California, Berkeley
John Hagood	The University of Utah
Charles J. Holland	Purdue University
William Hudson	Bowling Green State University
Barthel Huff	Queen's University
Werner Kohler	Virginia Polytechnic Institute and State University
F. Kozin	Polytechnic Institute of New York
Benny Levikson	Purdue University
James McKenna	Bell Laboratories
J. David Mason	The University of Utah
Steven Orey	The University of Minnesota
George C. Papanicolaou	Courant Institute of Mathematical Sciences
Mark A. Pinsky	Northwestern University
Ridgway Scott	Brookhaven National Laboratories

Marijean Seelback	NASA Ames Research Center
D. Stroock	University of Colorado
James Thomas	Colorado State University
Benjamin White	California Institute of Technology
Calvin Wilcox	The University of Utah
Randall Williams	California Institute of Technology
Ann Reed	Conference Secretary

Preface

This work is the collection of papers presented at a Conference on Stochastic Differential Equations and Applications held in Park City, Utah, February 17–20, 1976. The conference was sponsored by the Department of the Navy (Office of Naval Research) and the University of Utah.

The goal of the conference was to stimulate research in the theory and applications of stochastic differential equations by bringing together some of the mathematicians and applied scientists who are most active in the field.

The organizing committee consisted of J. David Mason and Calvin Wilcox of the University of Utah and George Papanicolaou of the Courant Institute of Mathematical Sciences.

We wish to thank the supporting agency and institution, the other members of the organizing committee, the participants and the authors, and Mrs. Ann Reed, who was the efficient conference secretary. Also, many thanks to Mrs. Ricki Rossi for her excellent work in typing this volume.

The Deterministic Itô-belated Integral is Equivalent to the Lebesgue Integral

BY

Richard B. Darst*

Abstract. Let $[a,b)$ be a bounded half open interval in R. Denote by $\mathcal{I}_b$ and $\mathcal{L}$ the sets of Itô-belated and Lebesgue integrable functions on $[a,b)$ to R. The purpose of this note is to substantiate the assertion in the title by showing that $\mathcal{I}_b$ is a subset of $\mathcal{L}$.

Because it is shown in [1] that $\mathcal{L} \subset \mathcal{I}_b$ it suffices to verify that $\mathcal{I}_b \subset \mathcal{L}$. To this end, we assume [1] (including notation) and suppose that $f \in \mathcal{I}_b - \mathcal{L}$. Then at least one of $f \vee 0$ and $f \wedge 0 \in \mathcal{I}_b - \mathcal{L}$, so we suppose that $f \geqslant 0$. Next we notice that if $f \wedge n$ were to belong to $\mathcal{L}$ for every positive integer n, then the inequalities $(L)\int_a^b f \wedge n = (I_b)\int_a^b f \wedge n \leqslant (I_b)\int_a^b f < \infty$ would be valid and f would belong to $\mathcal{L}$. Thus, we also suppose that $0 \leqslant f \leqslant n$ which implies that $0 \leqslant (I_b)\int_s^t f \leqslant n(t - s)$, $a \leqslant s \leqslant t \leqslant b$. Hence, the equation $F(t) = (I_b)\int_a^t f$ defines

* This note represents joint work with E. J. McShane.

a Lipschitz function on $[a,b]$, and there exists $g \in \mathcal{L}$ with $0 \leqslant g \leqslant n$ and $F(t) = (L)\int_a^t g = (I_b)\int_a^t g$. Denote $f - g$ by ϕ. Then $\phi \in \mathcal{I}_b - \mathcal{L}$, $|\phi| \leqslant n$ and $(I_b)\int_s^t \phi = 0$, $a \leqslant s \leqslant t \leqslant b$. Since $\phi \notin \mathcal{L}$, there exists $\varepsilon > 0$ such that $m^*(|\phi| > \varepsilon) > 8\varepsilon$. Multiplying by -1, and relabeling if necessary, suppose that $m^*(\phi > \varepsilon) > 4\varepsilon$. Remembering that $\phi \in \mathcal{I}_b$, choose (δ,δ^*) such that if π is a (δ,δ^*) (belated) partition of $[a,b)$, then $|S(\pi;\phi)| < \varepsilon^2$. Denote $(\phi > \varepsilon) \cap (\delta > 0)$ by E. From the Vitali covering $\{[x,x + \alpha];x \in E, \ 0 < \alpha < \delta(x)\}$ of E choose a finite sequence of pairwise disjoint elements $\{[x_i,x_i + \alpha_i]\}_{i\leqslant m}$ with $x_i < x_{i+1}$ and $\Sigma\alpha_i > 3\varepsilon$. On each complementary half open interval, $I_0 = [a,x_1)$, $I_j = [x_j + \alpha_j,x_{j+1})$, $I_m = [x_m + \alpha_m,b)$, choose a $(\delta,\delta^*/(m+1))$ partition π_j such that $|S(\pi_j,\phi)| < \varepsilon^2/(m+1)$. Let π be the (δ,δ^*) partition generated by the elements $(x_i,x_i+\alpha_i;x_i)$, $1 \leqslant i \leqslant n$, and the elements of π_j, $0 \leqslant j \leqslant m$. Thus, the contradiction $\varepsilon^2 > |S(\pi,\phi)| > \varepsilon\Sigma_i\alpha_i - \Sigma_j|S(\pi_j;\phi)| > 3\varepsilon^2 - \varepsilon^2$ obtains.

For the sake of completeness we mention that, since a function g of bounded variation on $[a,b)$ splits uniquely into a jump function and the difference of two mutually singular, continuous, non-decreasing functions, the analogous question for the case where g replaces the identity as integrator reduces to the case where g is continuous and non-decreasing. Using a slight twist of Vitali's theorem, our argument also applies to the latter case.

REFERENCES

1. E. J. McShane, *Stochastic Calculus and Stochastic Models*, Academic Press, New York (1974)

Department of Mathematics
Colorado State University
Fort Collins, Colorado

Optimal Stopping and Quasi Variational Inequalities

BY

Avner Friedman

1. Optimal Stopping and Variational Inequalities.

Let $w(t)$ be an n-dimensional Brownian motion and let

$$J_x(\tau) = E_x[\int_0^\tau e^{-\alpha t} f(w(t))dt + e^{-\alpha\tau}\phi(w(\tau))] \quad (\alpha > 0) \tag{1.1}$$

where f, ϕ are given functions and τ is a stopping time with respect to the Brownian motion. Consider the problem of finding

$$V(x) = \sup J_x(\tau) \tag{1.2}$$

and characterizing the time $\bar{\tau}$ for which

$$J_x(\bar{\tau}) = V(x).$$

Let us argue heuristically. At each point x, if we choose $\tau = 0$, that is, if we decide to stop, then we deduce

$$V(x) \geqslant \phi(x); \tag{1.3}$$

if however we choose $\tau > 0$, that is, if we decide to continue for time τ, and thereafter we proceed optimally, then we get

$$V(x) \geq E_x[\int_0^\tau e^{-\alpha t} f(w(t))dt + e^{-\alpha\tau} V(w(\tau))], \quad (1.4)$$

or, by Ito's formula,

$$0 \geq \int_0^\tau (\Delta V - \alpha V + f) w(t) dt. \quad (1.5)$$

Since τ is arbitrary,

$$\Delta V - \alpha V + f \leq 0. \quad (1.6)$$

Now, if the optimal thing to do is to stop, then equality should hold in (1.1). But, if the optimal thing to do is to continue until time $\hat{\tau}$, then equality should hold in (1.4) with $\tau = \hat{\tau}$. In view of (1.6), the latter equality implies $\Delta V - \alpha V + f = 0$ at $w(t)$, $0 \leq t \leq \hat{\tau}$, and, in particular, at x. We conclude

$$(V - \phi)(\Delta V - \alpha V + f) = 0. \quad (1.7)$$

Thus, V satisfies (1.3), (1.6), (1.7) or, in another notation,

$$\left.\begin{array}{l} (-\Delta V + \alpha V, w - V) \geq (f, w - V) \quad \text{for all} \quad w \geq \phi, \\ \\ V \geq \phi \end{array}\right\} \quad (1.8)$$

where $(\ ,\ )$ is the L^2 scalar product. Actually, in order not to worry about convergence at ∞, we may define $(\ ,\ )$ by

$$(\phi, \psi) = \int \phi\psi e^{-\mu|x|} dx$$

for some $\mu > 0$.

Conversely, one can show that if V is a solution of (1.8) which belongs to $C(R^n) \cap W^{2,2}_{loc}(R^n)$, then $V(x)$ satisfies (1.2) and the optional stopping time is the hitting time of the set $V = \phi$.

In order to complete the picture, we state the following theorem.

THEOREM 1.1. *(Under suitable assumptions on* f, ϕ*). There exists a solution* $V \in C(R^n) \cap W^{2,2}_{loc}(R^n)$ *of* (1.3).

For proof see [5], [8], [9], [10].

The system (1.8) is called an elliptic *variational inequality*. The above analysis can be extended to non-stationary problems with

$$J_{x,s}(\tau) = E_x[\int_s^\tau e^{-\alpha(t-s)} f(x(t),t)dt + e^{-\alpha(\tau-s)}\phi(w(\tau),\tau)I_{\tau<T} + e^{-\alpha(T-s)}h(w(T))I_{\tau=T}],$$

where the stopping times are restricted by $\tau \leqslant T$. This results in a parabolic variational inequality

$$\left(-\frac{\partial V}{\partial t} - \Delta V + \alpha V, w - V\right) \geqslant (f, w - V) \quad \text{for all} \quad w \geqslant \phi, \tag{1.9}$$

$$V \geqslant \phi; \qquad V(x,T) = h(x) \qquad (x \in R^n)$$

where $(\, , \,)$ is the L^2 scalar product in the (x,t)-space

2. ZERO SUM STOCHASTIC DIFFERENTIAL GAMES. Consider next

the "cost" function

$$J_x(\sigma,\tau) = E_x[\int_0^{\sigma\wedge\tau} e^{-\alpha t} f(w(t))dt + e^{-\alpha\sigma}\psi_1(w(\sigma))I_{\sigma<\tau} \tag{2.1}$$
$$+ e^{-\alpha\tau}\psi_2(w(\tau))I_{\tau\leq\sigma}],$$

where σ, τ are stopping times. This function arises from a game situation whereby a player 1 controls σ and attempts to minimize J_x, and a player 2 controls τ and attempts to maximize J_x. A *saddle point* is a pair (σ^*,τ^*) of stopping times such that

$$J_x(\sigma^*,\tau) \leqslant J_x(\sigma^*,\tau^*) \leqslant J_x(\sigma,\tau^*) \tag{2.2}$$

for all τ, σ. When such a saddle point exists, the number $V(x) = J_x(\sigma^*,\tau^*)$ is defined (independently of the saddle point) and it is called the *value of the game*.

Arguing heuristically we are led to the following inequalities for V.

$$(-\Delta V + \alpha V, w - V) \geqslant (f, w - V) \quad \text{for all} \quad w, \quad \psi_2 \leqslant w \leqslant \psi_1, \tag{2.3}$$

$$\psi_2 \leqslant V \leqslant \psi_1. \tag{2.4}$$

Naturally we must then a priori assume that $\psi_2 \leqslant \psi_1$. We can rewrite (2.3), (2.4) as follows

$$(-\Delta V + \alpha V + f, w - V) \geqslant 0 \quad \text{for all} \quad w \in K, \quad V \in K, \tag{2.5}$$

where K consists of all functions w satisfying $\psi_2 \leqslant w \leqslant \psi_1$.

The problem (2.5), with a general closed convex set K (in a suitable space), is called a *variational inequality*.

Again, there exists a solution $V \in C(R^n) \cap W^{2,2}_{loc}(R^n)$ of (2.3), (2.4), $V(x)$ is the value of the game, and a saddle point (σ^*,τ^*) is given by

$\sigma^*(\tau^*)$ is the hitting time of the set $V = \psi_1$ $(V = \psi_2)$.

For details see [9], [10], [16].

3. NON-ZERO SUM STOCHASTIC DIFFERENTIAL GAME. We now have two "costs":

$$J^1_x(\tau_1,\tau_2) = E_x\Big[\int_0^{\tau_1\wedge\tau_2} e^{-\alpha t} f_1(w(t))dt + e^{-\alpha\tau_1} I_{\tau_1<\tau_2}\phi_1(w(\tau_1)) + e^{-\alpha\tau_2} I_{\tau_1\geq\tau_2}\psi_1(w(\tau_2))\Big].$$

$$J^2_x(\tau_1,\tau_2) = E_x\Big[\int_0^{\tau_1\wedge\tau_2} e^{-\alpha t} f_2(w(t))dt + e^{-\alpha\tau_2} I_{\tau_2<\tau_1}\phi_2(w(\tau_2)) + e^{-\alpha\tau_1} I_{\tau_2\geq\tau_1}\psi_2(w(\tau_1))\Big]. \tag{3.1}$$

A *Nash equilibrium point* is a pair $(\hat\tau_1,\hat\tau_2)$ of stopping times such that

$$J^1_x(\hat\tau_1,\hat\tau_2) \leq J^1_x(\tau_1,\hat\tau_2),$$
$$J^2_x(\hat\tau_1,\hat\tau_2) \leq J^2_x(\hat\tau_1,\tau_2) \tag{3.2}$$

for all τ_1, τ_2.

Guided by the same heuristic considerations as before, we are led to the following system of inequalities for the functions

$$u_1(x) = J_x^1(\hat{\tau}_1,\hat{\tau}_2), \qquad u_2(x) = J_x^2(\hat{\tau}_1,\hat{\tau}_2).$$

$$u_1 \leqslant \phi_1, \qquad u_2 \leqslant \phi_2 \tag{3.3}$$

$$\begin{aligned} &\text{if } u_1(x) = \phi_1(x), \text{ then } u_2(x) = \psi_2(x); \\ &\text{if } u_2(x) = \phi_2(x), \text{ then } u_1(x) = \psi_1(x), \end{aligned} \tag{3.4}$$

$$\begin{aligned} &\Delta u_i + f_i \geqslant 0 \text{ on } \Sigma_i, \\ &(u_i - \phi_i)(\Delta u_i + f_i) = 0 \text{ on } \Sigma_i \quad (i = 1,2), \end{aligned} \tag{3.5}$$

where

$$\Sigma_1 = \{x : u_2(x) < \phi_2(x)\},$$

$$\Sigma_2 = \{x : u_1(x) < \phi_1(x)\}.$$

Letting

$$S_i = \{x : u_i = \phi_i\},$$

we can state the theorem.

THEOREM 3.1. *If* u_1, u_2 *form a "regular" solution of* (3.3)-(3.5), *then a Nash equilibrium point* $(\hat{\tau}_1,\hat{\tau}_2)$ *is given by* $\hat{\tau}_i$ = *hitting time of* S_i $(i = 1,2)$ *and* $u(x) = J_x^i(\hat{\tau}_1,\hat{\tau}_2)$.

This result is given in [4].

In order to formulate (3.3)-(3.5) in a more coherent form, introduce convex sets

$$K_1(u_2) = \{v \in W^{1,2}(R^n): v \leq \phi_1 \quad \text{a.e.},$$
$$u_2(x) \geq \phi_2(x) \quad \text{implies} \quad v(x) = \psi_1(x) \quad \text{a.e.}\},$$

$$K_2(u_1) = \{v \in W^{1,2}(R^n): v \leq \phi_2 \quad \text{a.e.},$$
$$u_1(x) \geq \phi_1(x) \quad \text{implies} \quad v(x) = \psi_2(x) \quad \text{a.e.}\}.$$

Consider the following problem for a pair (u_1,u_2):

$$u_1 \in K_1(u_2), \qquad u_2 \in K_2(u_1), \tag{3.6}$$

$$(\nabla_x u_1, \nabla_x(w - u_1)) + \alpha(u_1, w - u_1) \geq (f_1, w - u_1),$$
$$\text{for all} \quad w \in K_1(u_2),$$
$$(3.7)$$
$$(\nabla_x u_2, \nabla_x(w - u_2)) + \alpha(u_2, w - u_2) \geq (f_2, w - u_2),$$
$$\text{for all} \quad w \in K_2(u_1).$$

This problem is equivalent to (3.3)-(3.5).

Introducing

$$K(u) = \{v = (v_1,v_2): v_1 \in K_1(u_2), v_2 \in K_2(u_1)\},$$

$$\tilde{a}(u,v) = \sum_{i=1}^{2} [(\nabla_x u_i, \nabla_x v_i) + \alpha(u_i,v_i)],$$

where $u = (u_1,u_2)$, and letting $\tilde{f} = (f_1,f_2)$, we can rewrite (3.7) in the form

$$u \in K(u),$$

$$\tilde{a}(u, v - u) \geq (\tilde{f}, v - u). \tag{3.8}$$

This is actually a system of two sets of inequalities, one for u_1 and the other for u_2. But, the most important feature about this problem is that the convex set K depends on the solution u.

A variational inequality in which the convex set K depends on the solution u is called a *quasi variational inequality* (q.v.i.).

A standard procedure to prove existence of a solution is based on a fixed point theorem, i.e., for fixed u solve

$$w \in K(u),$$

$$\tilde{a}(w, v - w) \geq (\tilde{f}, v - w) \quad \text{for all} \quad w \in K(u),$$

and denote the solution w by Tu. Then prove that T has a fixed point. The difficulty in the present situation (and in most q.v.i.) is that T is not a continuous mapping.

THEOREM 3.2. *There exists a solution of* (3.8).

This was proved in [4]. The solution however is not known to be as "regular" as needed for Theorem 3.1.

4. OTHER QUASI VARIATIONAL INEQUALITIES.

Consider the process $x(t)$ given by

$$dx(t) = dw(t) + \sum_{i=1}^{N} \xi^i \delta(s - \theta^i),$$

where θ^i are stopping times, $\theta^1 < \theta^2 < \cdots < \theta^N < T$, ξ^i are non-negative random variables F_{θ^i} measurable, N is any finite number, and consider the cost function

$$J_{x,s}(\tau) = E_{x,s}[\int_s^T f(x(t),t)dt + N],$$

where $\tau = (\theta^1,\xi^1,\cdots,\theta^N,\xi^N)$ is the "impulse control". Let

$$u(x,t) = \inf_{\tau} J_{x,t}(\tau).$$

The heuristic principle of optimality (used already before) leads to the following conditions on u.

$$\left.\begin{aligned} & u \in K(u), \\ & \left(-\frac{\partial u}{\partial t} - \Delta u, w - u\right) \geqslant (f, w - u) \quad \text{for all} \quad w \in K(u) \end{aligned}\right\} \quad (4.1)$$

where

$$K(v) = \{w \in W^{1,2}(R^n) : w(x) \geqslant 1 + \inf_{\xi \geq 0} v(x + \xi)\}. \quad (4.2)$$

Here we then have another example of a q.v.i. This q.v.i. was introduced by Bensoussan and Lions [6] who were the first to introduce the concept of q.v.i.

If a solution of (4.2) is "regular" then the optimal impulse control τ is given as follows:

$$\left.\begin{array}{l} \theta^i \text{ is the hitting time of the set where} \\ \qquad u(x) = 1 + \inf_{\xi>0} u(x+\xi), \\ \text{and } \xi^i \text{ is a value of } \xi \text{ which solves this equation.} \end{array}\right\} \quad (4.3)$$

The existence of a solution is proved in [6]; it is not as "regular" as needed to ensure that the above rule (4.3) holds. A regular solution was established in [15]. More recent existence and regularity results on q.v.i. with $K(v)$ of the type (4.2) are given in [7]. See also [3] for a problem in hydraulics which is reduced to a q.v.i. of the form (4.1).

Another type of q.v.i. is when

$$u \geqslant \phi,$$

$$(-\Delta u, w - u) \geqslant (f, w - u) \quad \text{for all} \quad w \geqslant \phi,$$

(so that the set K is fixed) and f depends on the free boundary, i.e., on the boundary of the set where $u > \phi$. Examples of such q.v.i. arise in physics (see [13] for elliptic q.v.i. and [11], [12], [14] for parabolic q.v.i.). It would be interesting to find a probabilistic characterization for the solution of such q.v.i. The cost function should be non-Markovian.

5. A Quality Control Problem.

The results of this section are taken from [1], [2]. Let $\theta(t)$ be a Markov process with states $1,2,\cdots,n$ such that $p_{i,j}(t) = 0$ if $j < i$. With each state i we associate a product P^{λ_i}, an m-dimensional Brownian motion with drift λ_i. Let Ω be the space of m-vector functions $x(t)$, continuous on $0 \leq t < \infty$, and consider the process $(x(t),\theta(t))$ with measure $P^{\theta,x}$ which develops according to the random evolution of the P^{λ_i} as dictated by $\theta(t)$. More precisely, one can show that there exists a Markov process $P^{\theta,x}$ with respect to M_t, where $M_t = \hat{M}_{t+0}$ and

$$\hat{M}_t = \sigma\{(x(s),\theta(s)),\ 0 \leq s \leq t\},$$

such that $P^{\theta,x(s)}$ coincides with $P^{\lambda_i}_{x(s)}$ as long as $\theta(s) = i$.

Let $K_1,\cdots,K_{n-1}$, $c_1,\cdots,c_{n-1}$ be positive constants and let f be a function defined by $f(i) = c_{i-1}$, $(2 \leq i \leq n)$, $f(1) = 0$. Consider the costs

$$J^i_x(\tau) = E^{i,x}[K_i + \int_0^{\tau_1} f(\theta(s))ds$$

$$+I_{\theta(\tau_1)=i}[K_i+\int_{\tau_1}^{\tau_2} f(\theta(s))ds+I_{\theta(\tau_2)=i}[K_i+\int_{\tau_2}^{\tau_3} f(\theta(s))ds+\cdots]$$

$$+ I_{\theta(\tau_2)=i+1}[K_{i+1} + \int_{\tau_2}^{\tau_3} f(\theta(s))ds + \cdots]$$

$$+ \cdot \cdot \cdot \cdot + I_{\theta(\tau_2)=n-1}[K_{n-1} + \int_{\tau_2}^{\tau_3} f(\theta(s))ds + \cdots]]$$

$$+ I_{\theta(\tau_1)=i+1}[K_{i+1} + \int_{\tau_1}^{\tau_2} f(\theta(s))ds + I_{\theta(\tau_2)=i+1}[K_{i+1}$$

$$+ \int_{\tau_2}^{\tau_3} f(\theta(s))ds + \cdots] + I_{\theta(\tau_2)=i+2}[K_{i+2} + \int_{\tau_2}^{\tau_3} f(\theta(s))ds + \cdots]$$

$$+\cdots+ I_{\theta(\tau_2)=n-1}[K_{n-1} + \int_{\tau_2}^{\tau_3} f(\theta(s))ds + \cdots]]$$

$$+\cdots+I_{\theta(\tau_1)=n-1}[K_{n-1}+ \int_{\tau_1}^{\tau_2} f(\theta(s))ds+I_{\theta(\tau_2)=n-1}[K_{n-1}+ \int_{\tau_2}^{\tau_3} f(\theta(s))ds]].$$

Here, $E^{i,x}$ is the expectation associated with $P^{i,x}$ and $\tau = (\tau_1, \tau_2, \cdots)$ is an increasing sequence of "inspection times"; τ_{i+1} is an inspection time if it depends only on the information of F_t and of $\theta(\tau_i)$, where $\hat{F}_t = \sigma\{x(s),\ 0 \leq s \leq t\}$, $F_t = \hat{F}_{t+0}$.

The cost $J_x(\tau)$ represents a model of quality control problem (also called quickest detection problem). Each inspection costs K_i if θ at the preceding inspection was at state i. Another cost is due to the shift in θ from state ℓ to states $k \geq \ell$ during subsequent inspections; this cost is proportional to the time spent at the state k (with proportionality coefficient c_{k-1}). The problem is to find $\inf J_x(\tau)$ and characterize the optimal sequence of inspections.

Let

$$Q_i^{x,Y_1,\cdots,Y_{n-1}} = \sum_{j=i}^{n} \frac{Y_{j-i}}{1+Y_1+\cdots+Y_{n-i}} P^{j,x}, \quad (Y_0 \equiv 1), \tag{5.1}$$

and define processes

$$y_{i,j}(Y_1,\cdots,Y_{j-i},t) = \frac{Q_i^{x,Y_1,\cdots,Y_{n-i}}[\theta(t)=j|F_t]}{Q_i^{x,Y_1,\cdots,Y_{n-i}}[\theta(t)=i|F_t]}. \tag{5.2}$$

If $Y_1 = \cdots = Y_{j-i} = 0$, we get

$$y_{i,j}(t) = \frac{P^{i,x}[\theta(t)=j\,|\,F_t]}{P^{i,x}[\theta(t)=i\,|\,F_t]}. \qquad (5.3)$$

The process

$$(x(t), y_{i,i+1}(Y_1,t), \cdots, y_{i,n}(Y_1,\cdots,Y_{n-i},t))$$

with the measures (5.1) *form a Markov process.*

Its generator M_i can be computed explicitly. It is an elliptic operator defined for $x \in R^1$, $y_1 \geqslant 0, \cdots, y_{n-i} \geqslant 0$, and it is degenerate everywhere. When the λ_j are constants, the process $(y_{i,i+1}, \cdots, y_{i,n})$ forms by itself a Markov process. Its infinitesimal generator L_i is non-degenerate in the set $y_1 > 0, \cdots, y_{n-i} > 0$ if and only if the linear space spanned by the vectors

$$\lambda_{i+1} - \lambda_i, \cdots, \lambda_n - \lambda_i$$

is of dimension $n - i$. On the boundary, L_i is always degenerate.

Consider the following sequence of q.v.i.

$$V^i(x,y_1,\cdots,y_{n-i})$$

$$\leqslant K_i + \frac{V^i(x,0,\cdots,0)+y_1V^{i+1}(x,0,\cdots,0)+\cdots+y_{n-i-1}V^{n-1}(x,0)}{1+y_1+\cdots+y_{n-i}}, \qquad (5.4)$$

$$M_iV^i(x,y_1,\cdots,y_{n-i}) + \frac{c_1y_1 + c_{i+1}y_2 + \cdots + c_{n-1}y_{n-i}}{1+y_1+\cdots+y_{n-i}} \geqslant 0, \qquad (5.5)$$

$$[K_i + \frac{V^i(x,0,\cdots,0)+y_1V^{i+1}(x,0,\cdots,0)+y_{n-i-1}V^{n-1}(x,0)}{1 + y_1 + \cdots + y_{n-i}} - V^i(x,y_1,\cdots,y_{n-i})] \tag{5.6}$$

$$\cdot[M_iV^i(x,y_1,\cdots,y_{n-i}) + \frac{c_iy_1 + c_{i+1}y_2 + \cdots + c_{n-1}y_{n-i}}{1 + y_1 + \cdots + y_{n-i}}] = 0.$$

THEOREM 5.2. *Let* $V^1,\cdots,V^{n-1}$ *be "regular" solutions of the q.v.i.* (5.4)-(5.6). *Then*

$$\inf J_x^i(\tau) = V^i(x,0,\cdots,0) = J_x^i(\bar{\tau}^i),$$

where $\bar{\tau}^i = (\bar{\tau}_1^i,\bar{\tau}_2^i,\cdots)$ *and* $\bar{\tau}_{\ell+1}^i$ *is the following stopping rule. Knowing that* $\theta(\bar{\tau}_\ell^i) = j$, *proceed to make the next inspection when*

$$(x(t),y_{j,j+1}(t),\cdots,y_{j,n}(t))$$

hits the set where

$$V^j(x,y_1,\cdots,y_{n-j})$$

$$= K_j + \frac{V^j(x,0,\cdots,0)+y_1V^{j+1}(x,0,\cdots,0)+\cdots+y_{n-j-1}V^{n-1}(x,0)}{1 + y_1 + \cdots + y_{n-j}}$$

For $n = 2$ the q.v.i. can be solved fairly explicitly.

All the above extends to the case where P^{λ_i} are Poisson processes with parameters λ_i.

REFERENCES

1. R. F. Anderson and A. Friedman, *A quality contol problem and quasi variational inequalities*, to appear.
2. R. F. Anderson and A. Friedman, *Multi-dimensional quality control problems and quasi variational inequalities*, to appear.
3. C. Baiocchi, *Studio di un problema quasi-variozionale connesso a problemi di frontiera libera* , Bull. U.M.I. 4 12 (1975), 1-25.
4. A. Bensoussan and A. Friedman, *Non-zero sum stochastic differential games with stopping times and free boundary problems*, Trans. Amer. Math. Soc., to appear.
5. A. Bensoussan and J. L. Lions, *Problèmes de temps d'arrêt optimal et inéquations variationelles paraboliques*, Applicable Analys. 3 (1973), 267-295.
6. A. Bensoussan and J. L. Lions, *Nouvelles méthodes en contrôle impulsionnel*, Appl. Math. & Optimization 1 (1975), 289-312.
7. A. Bensoussan and J. L. Lions, *Sur le contrôle impulsionnel et les inéquations quasi variationnelles d'evolution*, C. R. Acad. Sci. Paris 280 (1975), 1049-1453.

8. H. Brezis, *Problèmes unilateraux*, J. Math. Pures Appl. 51 (1972), 1-168.

9. A. Friedman, *Stochastic differential games and variational inequalities*, Arch. Rat. Mech. Anal. 51 (1973), 321-346.

10. A. Friedman, *Regularity theorems for variational inequalities in unbounded domains and applications to stopping time problems*, Arch. Rat. Mech. Anal. 52 (1973), 134-160.

11. A. Friedman, *A class of parabolic quasi-variational inequalities II*, J. Diff. Eqs., to appear.

12. A. Friedman and R. Jensen, *A parabolic quasi-variational inequality arising in hydraulics*, Ann. Scu. Norm. Sup. Pisa, 2 (4) (1975), 421-468.

13. A. Friedman and R. Jensen, *Elliptic quasi-variational inequalities and application to a non-stationary problem in hydraulics*, Ann. Scu. Norm. Sup. Pisa, 3 (4) (1976).

14. A. Friedman and D. Kinderlehrer, *A class of parabolic quasi-variational inequalities*, J. Diff. Eqs., to appear.

15. J. L. Joly, U. Mosco and G. Troianjello, *Un résultat de régularité pour inéquation quasi-variationnelle intervenant dans un problème de contrôle impulsionnel*, C. R. Acad. Sci. Paris, 279 (1974), 937-940.

16. N. U. Krylov, *Control of Markov processes and W-spaces*, Math. USSR-Izv., 5 (1971), 233-266 [Izv. Akad. Nauk SSSR,

Ser. Math., 35 (1971), 224-255].

Department of Mathematics
Northwestern University
Evanston, Illinois

This paper is partially supported by National Science Foundation Grant MPS 72-04959 A02.

THE REGULAR EXPANSION IN SINGULARLY PERTURBED ELLIPTIC EQUATIONS

BY

CHARLES J. HOLLAND

1. INTRODUCTION. Recently we have used probabilistic methods to derive some of the regular and ordinary boundary layer expansions for certain classes of singularly perturbed second order elliptic partial differential equations subject to Dirichlet and Neumann boundary conditions. This talk will survey some of these results and indicate some extensions of the regular expansion for the Neumann and mixed boundary value problems. The probabilistic method for deriving these expansions depends upon the representation of the solution of the partial differential equation as the expected value of a functional of an Ito stochastic differential equation. See Fleming [3] and Freidlin [4] for excellent surveys of the previous uses of this approach in deriving expansions in singular perturbation problems.

For ease of exposition we will state some results in less generality than we or others have actually proven them. Thus let us consider the linear elliptic partial differential equation in two independent variables $x = (x_1 , x_2)$

$$\varepsilon(\Delta_x U) + A(x)U_{x_1} + B(x)U_{x_2} + C(x)U + f(x) = 0 , C < 0 , \quad (1)$$

in some domain G satisfying the boundary data

$$\alpha(x)U(x) + \beta(x) \frac{\partial U}{\partial n}(x) = g(x) \quad (2)$$

on ∂G. $\frac{\partial U}{\partial n}$ denotes the derivative in the direction of the exterior normal to ∂G, and $\alpha(x) \geqslant 0$, $\beta(x) \geqslant 0$ and $\alpha(x)^2 + \beta(x)^2 = 1$.

Throughout let us assume that the functions A, B, C, f, g are infinitely differentiable in some domain D containing $G \cup \partial G$. We shall assume that the boundary of G consists of a finite number of closed curves; the coordinates x_1 and x_2 are each infinitely differentiable as a function of arc length. Finally there exists $\overline{C} < 0$ such that $C(x) < \overline{C}$ for all x in G. We shall also assume that the boundary data (2) is such that (1) has a classical solution for each $\varepsilon > 0$.

The outline of the talk is as follows. We discuss in Section II the regular expansion for Dirichlet data and in Section III the regular expansion for zero Neumann data. For the Neumann case we require a more stringent behavior on the boundary than for the Dirichlet case of the characteristics of the reduced equation (1) with $\varepsilon = 0$. We first state in

Theorem 2 our result which was proved in [8]. Then in Theorem 3 we change the requirement on the boundary behavior that was imposed in [8]. Here we require that locally all characteristics intersect the boundary in the direction of the exterior normal to the boundary. We give an example to show that either the stringent behavior in Theorem 2 or 3 is necessary to derive the regular expansion. This example shows that near a "stagnation" point which lies between two "sinks" one can expect convergence of the solution U^ε of (1), (2) to an appropriate solution of the reduced equation although a regular expansion does not hold there. We discuss the case of mixed boundary data in Section IV. Our method consists in reducing the problem to one for which some modification of the results in Sections II and III is applicable. Finally, in Section V we mention some other work on probabilistic methods in singularly perturbed elliptic partial differential equation and some extensions of the work presented here.

We are interested in determining the behavior as $\varepsilon \downarrow 0$ of the solution to equation (1). The asymptotic behavior depends upon the characteristics of the reduced equation (1) with $\varepsilon = 0$ and the boundary conditions. In certain subsets R of the domain G one would hope to approximate U^ε by a regular expansion in powers of ε:

$$U^\varepsilon = U^0 + \varepsilon\theta_1 + \varepsilon^2\theta_2 + \cdots + \varepsilon^k\theta_k + o(\varepsilon^k) \quad . \tag{3}$$

Equations for the functions U^o, $\theta_1, \theta_2, \ldots$ would be found by formally substituting (3) into (1). Thus U^o satisfies the reduced equation

$$AU^o_{x_1} + BU^o_{x_2} + CU^o + f = 0 \tag{4}$$

while θ_j satisfy (with $\theta_o = U^o$) the equations

$$A(\theta_j)_{x_1} + B(\theta_j)_{x_2} + C\theta_j + \Delta_x\theta_{j-1} = 0 . \tag{5}$$

The functions U^o, θ_j satisfy appropriate boundary data along a certain portion of the boundary of $\overline{\partial G \cap \partial R}$.

Let us now discuss the subsets of G in which the regular expansion is established. For $(x_1, x_2) \in G$ the *characteristic* of (4) is the solution $\xi_x(t)$, $t > 0$, of the differential equation

$$\frac{d\xi_1}{dt}(t) = A(\xi_1(t), \xi_2(t)), \quad \xi_1(0) = x_1,$$

$$\frac{d\xi_2}{dt}(t) = B(\xi_1(t), \xi_2(t)), \quad \xi_2(0) = x_2 .$$

A region $R \subset \overline{G}$ is called a *type I acceptable region* if it has the following properties:

(i) There exists a continuous one-to-one function $h(s) = (h_1(s), h_2(s))$ on $[0, 1]$ with values in $G \cap R$.

(ii) For $x = h(s)$, $0 \leqslant s \leqslant 1$ there exists $\tau_x > 0$ such that $\xi_x(t) \in G \cap R$ for $0 < t < \tau_x$ and $\xi_x(\tau_x) \in \partial G$.

(iii) $R = \{\xi_x(t)\colon\ x = h(s),\ 0 \leqslant s \leqslant 1;\ 0 \leqslant t \leqslant \tau_x\}$.

(iv) Let $\Gamma(R) = \{ \xi_x(\tau_x): x = h(s), 0 \leqslant s \leqslant 1\}$. If $x \in \Gamma(R)$, then $(A(x), B(x)) \cdot n(x) \neq 0$, where $n(x)$ is the exterior normal to G at x.

(v) There exists a unique point $y \in \Gamma(R)$ such that $(A(y), B(y))$ is parallel to $n(y)$. If $x \in \Gamma(R)$, $x \neq y$, and x is in the counterclockwise (clockwise) direction from y, then the vector $(A(x), B(x))$ is in a clockwise (counterclockwise) direction from $n(x)$.

If $G \cap R$ in (i) is replaced by $\partial G \cap R$ then we call R a *type II acceptable region.* If condition (v) in the definition is deleted, then we call R a *type I semi-acceptable region (type II semi-acceptable region)* if $G \cap R$ $(\partial G \cap R)$ is used in (i). In type II regions we use the notation $\Lambda(R) = \{h(s): 0 \leqslant s \leqslant 1\}$. It will be convenient for stating results to assume that $\Lambda(R) \subset \{ x_1 = 0\}$.

2. The Dirichlet Case.

We first discuss the case when Dirichlet boundary data $U = g$ is given on the portion of the boundary $\Gamma(R)$ of a type I semi-acceptable region.

THEOREM 1. (a) *Suppose that* R, R' *are type I semi-acceptable regions with* $R \subset R'$ *and* $\partial R \cap (\partial R' - \partial G) = \phi$ *and that Dirichlet data* $U = g$ *is given on* $\Gamma(R')$. *Then for any positive integer* k *the regular expansion* (3) *holds uni-*

formly on R. *The functions* $U^0, \theta_1, \theta_2, \cdots$ *are the solutions of the equations* (4) *and* (5) *taking the boundary* $U^0 = g$ *and* $\theta_j = 0$ *on* $\Gamma(R')$.

(b) *Suppose that* R, R' *are type II semi-acceptable regions and Dirichlet data* $U = g$ *is given on* $\Gamma(R')$ *and* $\Lambda(R')$. *Then the regular expansion* (3) *holds uniformly on* $R \cap \{x_1 \geqslant \delta(\varepsilon)\}$, *where* $\delta(\varepsilon) = \varepsilon^m$ *for any fixed* m *satisfying* $0 < m < \frac{1}{2}$.

Theorem 1 can in fact be proven by probabilistic methods for a wide class of semilinear equations which includes (1). See Fleming [3] for the proof of part (a) and Holland [7] for the proof of part (b). The minor extension part (b) is necessary in order to establish the ordinary boundary layer expansion which arises along $\Lambda(R)$. This boundary layer occurs because in general the solution U^0 does not match the boundary data given on $\Lambda(R)$.

We remark that the assumption $C < 0$ is not necessary, it can be eliminated provided there is an a priori bound $|U^\varepsilon| \leqslant K$ on R', $0 < \varepsilon < \varepsilon_0$.

3. The Zero Neumann Data Case.

The following theorem was proved in [8].

THEOREM 2. *Let* R,R' *be type I acceptable regions with* $R \subset R'$ *and* $\partial R \cap (\partial R' - \partial G) = \phi$ *and let* $R^* \subset G$ *be* *an open convex set with boundary of class* C^3 *such that* $R' \subset \overline{R}^*$. *Assume that zero Neumann data is given on* $\Gamma(R')$. *Assume further that* $U^0, \theta_1, \theta_2, \cdots$ *are solutions of equations* (4) *and* (5) *in* R' *of class* C^3 *satisfying the boundary data* $\frac{\partial U^0}{\partial n} = \frac{\partial \theta_j}{\partial n} = 0$ *on* $\Gamma(R')$. *Then for any positive integer* k, *the expansion* (3) *holds uniformly on* R.

OUTLINE OF PROOF. See the proof of Theorem 3 which uses the basic ideas needed for this proof.

A careful reading of the proof in [8] shows that we also have the following.

COROLLARY 1. *Suppose that* R, R' *are type II acceptable regions with zero Neumann data given on* $\Gamma(R')$ *and* $\Lambda(R')$. *Assume the existence of* R^* *and* U^0 *as in Theorem* 2. *Then* $U^\varepsilon \to U^0$ *uniformly on* R.

Unlike the case of Dirichlet data on $\Lambda(R')$, no boundary layer occurs in the zeroth order expansion.

To prove the regular expansion in the Neumann problem we required the additional property (v) on the behavior of the characteristics crossing the boundary $\Gamma(R)$. At the point y as required in (v), the reduced equation (2) and the boundary condition allowed us to determine the value U^0 of the solution. Let us now discuss the behavior of the solution U^ε in type I semi-acceptable regions where property (v) is not satisfied.

We first consider an example.

EXAMPLE 1. Let the domain G be defined by

$$G = \{(x_1, x_2): 0 \leqslant x_1 \leqslant 2, \quad -2 \leqslant x_2 \leqslant 2\} \cup$$
$$\{(x_1, x_2): (x_1 - 1)^2 + (x_2 - 2)^2 \leqslant 1\} \cup$$
$$\{(x_1, x_2): (x_1 - 1)^2 + (x_2 + 2)^2 \leqslant 1\} ,$$

and consider equation (1) with $A = 1$ and $B = x_2^3(1-x_2)(1+x_2)$. Now for any k with $0 < k < 2$

$$D_1 = \{\xi_x(t): 0 \leqslant t \leqslant \tau_x, \ x = (k, s), -2 \leqslant s \leqslant 2\}$$

is a semi-acceptable region, but is not an acceptable region. Note that $\Gamma(D_1)$ contains three points y for which $(A(y), B(y))$ is parallel to $n(y)$. Now the regular expansion for the Dirichlet problem is valid on D_1, but for general C and f the regular expansion for the Neumann problem is not. However, for k as before and any $\delta > 0$,

$$D_2 = \{\xi_x(t): 0 \leqslant t \leqslant \tau_x, \ x = (k, s), \delta \leqslant s \leqslant 2\}$$

and

$$D_3 = \{\xi_x(t): 0 \leqslant t \leqslant \tau_x, \ x = (k, s), -2 \leqslant s \leqslant -\delta\}$$

are both type I acceptable regions and the regular expansion is valid on D_2 and on D_3 when zero Neumann data is specified on $\{x_1 = 2\} \cap \partial G$. Let U_2^0 and U_3^0 be the functions in the expansion (3) for the acceptable regions D_2 and D_3 . The function U_2^0 is partially determined from the condition at the "sink" (2, 1)

$$C(2, 1)U_2^0(2, 1) + f(2, 1) = 0 ,$$

and U_3^0 is partially determined from the condition at the "sink" (2, -1)

$$C(2, -1)U_3^0(2, -1) + f(2, -1) = 0 .$$

Given the construction as described in [8] of the functions U_2^0, U_3^0, it is easy to check that for each x_1 in $(k, 2)$,

$$\lim_{c \downarrow 0} U_2^0(x_1, c) \quad \text{and} \quad \lim_{c \uparrow 0} U_3^0(x_1, c)$$

exist and are equal. In fact, the limit is given by the solution $U^*(x_1, 0)$ to the ordinary differential equation

$$U^*_{x_1}(x_1, 0) + C(x_1, 0)U^*(x_1, 0) + f(x_1, 0) = 0$$

with boundary condition at the "stagnation" point

$U^*_{x_1}(2, 0) = 0$.

Let

$$U(x) = \left\{ \begin{array}{l} U_2^0(x_1, x_2),\ x_2 > 0 \\ U^*(x_1, 0),\ x_2 = 0 \\ U_3^0(x_1, x_2),\ x_2 < 0 \end{array} \right\} .$$

Then $U(x)$ is continuous in D_1, but in general is not differentiable along $x_2 = 0$. It can be shown that $U^\varepsilon \to U^0$ in D_1, although no expansion of the form (3) with $k \geqslant 1$ is valid in a neighborhood of $x_2 = 0$.

It is possible to formulate a general result about convergence of U^ε as $\varepsilon \downarrow 0$ for domains of the type discussed in this example, however, we choose not to do so.

We now turn our attention to the case when property (v) in the definition of an acceptable region R is replaced by the condition

(v') $(A(y), B(y))$ is parallel to $n(y)$ for each point $y \in \Gamma(R)$.

THEOREM 3 AND COROLLARY 2. *Let property* (v) *be replaced by property* (v') *in the regions* R, R' *in the statements of Theorem* 2 *and Corollary* 1. *Then the conclusions of Theorem* 2 *and Corollary* 1 *remain valid.*

OUTLINE OF PROOF. The proof requires some alterations in the proof of Theorem 2 presented in [8]. Let us outline the approach by establishing convergence of U^{ε} to U^{0} in R.

We make a probabilistic representation of the solution $U^{\varepsilon}(x)$. Consider for $\varepsilon > 0$ and for $x \in R^*$ the process $\xi_x^{\varepsilon}(t)$ which is the solution of the stochastic differential equation

$$d\xi_x^{\varepsilon}(t) = \begin{pmatrix} A(\xi_x^{\varepsilon}(t)) \\ \\ B(\xi_x^{\varepsilon}(t)) \end{pmatrix} dt + (2\varepsilon)^{1/2}\, dw, \tag{7}$$

$\xi_x^{\varepsilon}(0) = x$, where w is 2 dimensional Brownian motion, with reflection of the process in the interior normal direction of the boundary of R^*. The existence and uniqueness of such a process was proved in Theorem 2.1 of Bensoussan-Lions [2]. We have chosen reflection at the boundary of R^*, since R^* is convex, which will allow us to use some estimates in [2] of process $\xi_x^{\varepsilon}(t)$.

For x in $\overline{R}$, let γ_x^{ε} be the first time the process $\xi_x^{\varepsilon}(t)$ reaches the closed set $\overline{\partial R' - \partial G}$, and let $\tau_x^{\varepsilon} = \min(\gamma_x^{\varepsilon}, \varepsilon^{-1/6})$. Then, for x in $\overline{R}$, U^{ε} has the probabilistic representation

$$U^{\varepsilon}(x) = E\int_0^{\tau_x^{\varepsilon}} [\exp\int_0^{t} C(\xi_x^{\varepsilon}(s))ds] f(\xi_x^{\varepsilon}(t))dt + E[\exp\int_0^{\tau_x^{\varepsilon}} C(\xi_x^{\varepsilon}(s))ds]\cdot U^{\varepsilon}(\xi_x^{\varepsilon}(\tau_x^{\varepsilon})). \tag{8}$$

We also make a representation of the solution $U^o(x)$. Denote $\Gamma(R')$ by means of Γ . For $x \in \Gamma$, define

$$\pi \begin{pmatrix} A(x) \\ B(x) \end{pmatrix} = \begin{pmatrix} A(x) \\ B(x) \end{pmatrix} - [(A(x), B(x)) \cdot n(x)] n(x) ,$$

where $n(x)$ is the exterior normal to G. For $x \in R'$, let $\xi_x^0(t)$ be the solution to

$$\frac{d\xi_x^0}{dt}(t) = (1 - \chi_\Gamma(\xi_x^0(t))) \cdot \begin{pmatrix} A(\xi_x^0(t)) \\ B(\xi_x^0(t)) \end{pmatrix} + \chi_\Gamma(\xi_x^0(t)) \pi \begin{pmatrix} A(\xi_x^0(t)) \\ B(\xi_x^0(t)) \end{pmatrix} \tag{9}$$

with initial condition $\xi_x^0(0) = x$. In (9), χ_Γ is the indicator function of the set Γ . The equation (9) is also discussed in Bensoussan-Lions [2]. Recall the properties of the acceptable region R' . Note that if $x \in \overline{R}$, once $\xi_x^0(t)$ reaches the set Γ, then $\xi_x^0(t)$ is fixed at that point on Γ. Then, for any constant $q > 0$, U^o satisfies

$$U^o(x) = \int_0^q \exp \int_0^t c(\xi_x^0(s))ds \cdot f(\xi_x^0(t))dt + \exp \int_0^q c(\xi_x^0(s))ds \, U^o(\xi_x^0(q)) .$$

Replacing q with τ_x^ε, we obtain that for any $\varepsilon > 0$,

$$U(x) = E\int_0^{\tau_x^\varepsilon}[\exp \int_0^t c(\xi_x^0(s))ds] \cdot f(\xi_x^0(t))dt$$
$$+ \; E[\exp \int_0^{\tau_x^\varepsilon} c(\xi_x^0(s))ds] \cdot U(\xi_x^0(\tau_x^\varepsilon)) \; . \tag{10}$$

We show that $U^\varepsilon \to U^0$ through use of the representations (8) and (10).

To do this we need to derive some probabilistic estimates on the behavior of the process $\xi_x^\varepsilon(t)$. For functions h defined on $[0, t]$, let $\|h\|_t = \sup_{0 < t' < t} |h(t')|$.

Now let $\delta, T > 1$ be such that if $x \in \overline{R}$ and

$$\|\xi_x^\varepsilon - \xi_x^0\|_T < \delta,$$

then $\xi_x^\varepsilon(T) \in \Gamma$ and $\xi_x^\varepsilon(t)$, $0 \leq t \leq T$, never contacts the set $\overline{\partial R' - \partial G}$.

Now if

$$\|(2\varepsilon)^{1/2} \int_0^t (\xi_x^\varepsilon(s) - \xi_x^0(s), dw(s))\|_T \leq C(\varepsilon), \tag{11}$$

then

$$\|\xi_x^\varepsilon - \xi_x^0\|_T \leq \sqrt{(4\varepsilon T + C(\varepsilon))\exp MT} \tag{12}$$

uniformly for any initial condition x in $\overline{R}$. In (12) M is a Lipschitz constant for the vector (A,B).

Choose $C(\varepsilon) = \varepsilon^{5/12}$ and $\varepsilon^* > 0$ so that for $\varepsilon < \varepsilon^*$

the right side of (12) is less than $\delta(1 + T^{-1}\varepsilon^{-1/6})^{-1}$. (This represents a change in the proof of Theorem 2 in [8].)

Repeating the calculations in the proof of Theorem 2 in [8] we obtain that there exists a positive constant β such that

$$Pr\{\tau_x^\varepsilon \neq \gamma_x^\varepsilon\} \geq 1 - 4\exp(2\beta T)\exp(\tfrac{-\varepsilon^{-1/12}}{2})(1 + T^{-1}\varepsilon^{-1/6})$$

and therefore $Pr\{\tau_x^\varepsilon \neq \gamma_x^\varepsilon\} \to 1$ as $\varepsilon \to 0$ uniformly for x in $\overline{R}'$. With these estimates we obtain that the last term on the right side of equation (8) and of equation (10) have limit zero as $\varepsilon \downarrow 0$.

The convergence of the first term on the right side of (8) to the first term on the right side of (10) follows exactly as in [8]. This concludes the outline of the proof.

4. Mixed Boundary Value Problems.

Let us now discuss briefly problems when other than Dirichlet or zero Neumann data is specified on the set $\partial G \cap \overline{R}$. Essentially, our method consists in making changes of variables to reduce the given problem to one of the form we can apply the results of the earlier sections with some modifications.

Suppose that R, R', R^* are as in Theorem 2 and that $\beta(x) \neq 0$ on $\Gamma(R')$. We wish to derive the appropriate regular expansion (3) on R . We assume the following:

(D) there exists a function γ of class $C^\infty(R')$ such that $\frac{\partial\gamma(x)}{\partial n} = \frac{\alpha(x)}{\beta(x)}$ on $\Gamma(R')$, and that there exists a function H of class $C^{\infty}(R')$ such that $\frac{\partial H(x)}{\partial n} = \frac{g(x)}{\beta(x)} e^{\gamma(x)}$ on $\Gamma(R')$.

Now, for $\varepsilon > 0$, consider the function

$$W^{\varepsilon}(x) = U^{\varepsilon}(x) \; e^{\gamma(x)} - H(x)$$

which satisfies the boundary condition $\frac{\partial W^{\varepsilon}}{\partial n}(x) = 0$ on $\Gamma(R')$. For each $\varepsilon > 0$, W^{ε} satisfies the equation

$$\varepsilon\Delta_x W + A(x,\varepsilon)W_{x_1} + B(x,\varepsilon)W_{x_2} + C(x,\varepsilon)W + f(x,\varepsilon) = 0 \quad (13)$$

in R, where we have defined

$$A(x,\varepsilon) = A(x) - 2\varepsilon \frac{\partial\gamma}{\partial x_1}(x),$$

$$B(x,\varepsilon) = B(x) - 2\varepsilon \frac{\partial\gamma}{\partial x_2}(x),$$

$$C(x,\varepsilon) = C(x) - A(x)\frac{\partial\gamma}{\partial x_1}(x) - B(x)\frac{\partial\gamma}{\partial x_2}(x) - \varepsilon\Delta_x\gamma(x) + \varepsilon\left\{\left[\frac{\partial\gamma}{\partial x_1}(x)\right]^2 + \left[\frac{\partial\gamma}{\partial x_2}(x)\right]^2\right\}$$

and

$$\begin{aligned} f(x,\varepsilon) = {} & f(x)e^{\gamma(x)} + C(x)H(x) + A(x)H_{x_1}(x) \\ & + B(x)H_{x_2}(x) - A(x)\gamma_{x_1}(x)H(x) - B(x)\gamma_{x_2}(x)H(x) \\ & + \varepsilon\Big[\Delta_x H(x) - H(x)\Delta_x\gamma(x) \\ & \quad + H(x)\left\{\left[\frac{\partial\gamma}{\partial x_1}(x)\right]^2 + \left[\frac{\partial\gamma}{\partial x_2}(x)\right]^2\right\}. \end{aligned}$$

Let us compute a formal regular expansion for W^{ε}:

$$W^\varepsilon(x) = W^o(x) + \sum_{j=0}^{k} \varepsilon^j Y_j(x) + o(\varepsilon^k) . \tag{14}$$

We find that $W^o(= Y_o)$ satisfies

$$A(x)W^o_{x_1} + B(x)W^o_{x_2} + \left[C(x) - A(x)\frac{\partial\gamma}{\partial x_1}(x) - B(x)\frac{\partial\gamma}{\partial x_2}(x) \right] W^o$$
$$+ f(x, 0) = 0 , \tag{15}$$

and Y_j satisfies

$$A(x)(Y_j)_{x_1} + B(x)(Y_j)_{x_2} + C(x)Y_j + \Delta_x(Y_{j-1})$$
$$-2(\frac{\partial\gamma}{\partial x_1}(x))(Y_{j-1})_{x_1} - 2(\frac{\partial\gamma}{\partial x_2}(x))(Y_{j-1})_{x_2} \tag{16}$$
$$-(\Delta_x\gamma(x))Y_{j-1} + \left[(\frac{\partial\gamma}{\partial x_1})^2 + (\frac{\partial\gamma}{\partial x_2})^2 \right] Y_{j-1}$$
$$+ [j = 1] \Bigg\{ \Delta_x H(x) - H(x)\Delta_x\gamma(x)$$
$$+H(x)\left\{ \left[\frac{\partial\gamma}{\partial x_1}(x) \right]^2 + \left[\frac{\partial\gamma}{\partial x_2}(x) \right]^2 \right\} \Bigg\} = 0.$$

Above $[j = 1] = 1$ if $j = 1$ and $= 0$ if $j \neq 1$. The boundary conditions are

$$\frac{\partial W^o}{\partial n} = \frac{\partial Y_j}{\partial n} = 0 \tag{17}$$

We would now like to apply the methods of proof of Theorems 2 and 3 to derive the regular expansion for W^ε. The equation (13) is similar to (1), except now the coefficients of (13) depend upon ε in a smooth manner. This causes no problems. However, we no longer know that the coefficient $C(x, \varepsilon)$ of W is negative in R'. This complicates matters slightly, but we can easily show that

$$(A, B) \cdot \left(\frac{\partial \gamma}{\partial x_1}, \frac{\partial \gamma}{\partial x_2}\right) \geqslant 0$$

at the point y on $\Gamma(R)$ and hence $C(x, \varepsilon) < 0$ in some neighborhood of y for sufficiently small $\varepsilon > 0$. A careful examination of the proof of Theorems 2 and 3 show that this is sufficient in order to establish the expansion. This follows since, loosely speaking, the reflecting process $\xi_x^\varepsilon(t)$ reaches a neighborhood of y in finite time and then spends "most" of the time near the point y where $C(x, \varepsilon) < 0$. We can thus exploit the exponential decay in the corresponding term in (8) involving $C(x, \varepsilon)$. We thus have the following result for W^ε which yields an expansion for U^ε.

THEOREM 4. (a) *Suppose that* R, R', R^* *are as in Theorem* 2 *and that assumption* (D) *is satisfied. Suppose that* $W^0, Y_1, Y_2, \cdots$ *are solutions of class* $C^3(\overline{R}')$ *satisfying the boundary data* (17). *Then for any positive integer* k *the expansion* (13) *holds uniformly on* R.

(b) *Part* (a) *is true if* R, R', R^* *are as in Theorem* 3.

If an expansion for W^ε is valid, hence for U^ε also, then it is not necessary to first compute the expansion for

W^ε. Instead, one directly seeks an expansion (3) for U^ε. The functions $U^o, \theta_1, \theta_2, \cdots$ satisfy the equations (4), (5) with the boundary condition (2) on R'.

5. CONCLUDING REMARKS. The results presented have extensions to the case of $n \geqslant 2$ independent variables. The extensions of Theorems 1 and 3 are immediate. Theorem 2 requires that the assumption (v) be replaced by the assumption that the corresponding n dimensional boundary process obtained from (9) is such that the point y is asymptotically stable and paths of (9) starting in $\Gamma(R)$ remain in $\Gamma(R)$.

We derived the ordinary boundary layer expansion in semi-acceptable regions when Dirichlet data is given on $\Lambda(R)$. Elsewhere we will treat the case of ordinary boundary layer expansions when Neumann or mixed data is specified on $\Lambda(R)$.

Another question of interest for equation (1) has been the case when all of the directed characteristics starting in G remain in G for all $t > 0$. The case of Dirichlet data was first treated by Ventsel and Freidlin [9]. This problem has also been treated by Friedman [6]. Recently Anderson and Orey [1] have treated the case when $\frac{\partial U}{\partial n} = g$ is given on the boundary of the domain G. All of the above papers use probabilistic methods.

The assumption $C \leqslant 0$ eliminates some interesting phenomena. It was this assumption that allowed the "neglect" of the last terms on the right side of (8) and (10). Freidlin

[5] considered the case when $C = 0$ under certain assumptions. An interesting case occurs when for an acceptable region R zero Neumann data is given on $\Gamma(R)$, but Dirichlet data is given on some segments of the boundary of the domain G connected to $\Gamma(R)$.

REFERENCES

1. R. Anderson and S. Orey, *Small random perturbations of dynamical systems with reflecting boundary conditions*, to appear.

2. A. Bensoussan and J. L. Lions, *Diffusion processes in bounded domains and singular perturbation problems for variational inequalities with Neumann boundary conditions*, Probabilistic methods in differential equations, Springer Verlag Lecture Notes in Mathematics #451, 8-25.

3. W. Fleming, *Stochastically perturbed dynamical systems*, Rocky Mountain J. of Math., 4(1974), 407-433.

4. M. Freidlin, *Markov processes and differential equations*, pp. 1-55, Progress in Mathematics, Volume 3, editor R. V. Gamkrelidze, Plenum Press, N.Y. 1969.

5. __________, *A Mixed Boundary Value Problem for Elliptic Differential Equations of Second Order with a Small Parameter*, Dokl. Akad. Nauk SSSR, 143(1962), 1300-1303 Soviet Math Dokl., 3(1962), 616-620.

6. A. Friedman, *Small random perturbation of dynamical systems and applications to parabolic equations*, Ind. U. Math. J., 24 (1974), 533-553, 24 (1975), 903.

7. C. J. Holland, *Singular perturbations in elliptic partial differential equations*, J. Differential Equations, 20 (1976), 247-265.

8. ____________, *The regular expansion in the Neumann problem for elliptic equations*, Communications in Partial Differential Equations, 1 (1976), 191-213.

9. A. D. Ventcel and M. Freidlin, *On small random perturbations of dynamical systems*, Russ. Math. Surveys, 25 (1970) No. 1, 1-56 (Uspekhi Math. Nauk, 28 (1970), No. 1, 3-55).

Department of Mathematics
Purdue University
West Lafayette, Indiana

FLUCTUATION PHENOMENA IN UNDERWATER SOUND PROPAGATION, I.

BY

WERNER KOHLER

AND

GEORGE C. PAPANICOLAOU*

ABSTRACT. The study of long-range acoustic propagation in the ocean is basically a study of wave propagation in a random medium. The complexity of the problem precludes an exact analysis; a number of heuristically plausible simplifications must be made to make the analysis tractable. In this paper we shall establish the foundation for this analysis by considering more basic problems of random wave propagation. The simplest of such problems, the source-excited one dimensional slab, will be discussed at length. The more difficult problem of propagation in a randomly perturbed waveguide will then be considered and the required approximations discussed. A second paper [1] will address the specific problem of underwater sound propagation.

1. INTRODUCTION.

Problems involving wave

---

* This work was partially completed at the Applied Mathematics Summer Institute, 1975, which is supported by the ONR under Contract No. N00014-75-C-0921.

propagation in a random medium arise in a number of physically important contexts. Increasing demands upon electromagnetic communications systems, for example, are forcing carrier frequencies upward into the millimeter wavelength and optical regimes (cf. [2], [3]). At such frequencies, slight misalignments, imperfections, etc. become quite important; their unpredictability necessitates a stochastic modeling. Similarly, the effects of atmospheric turbulence upon laser beam propagation [4], [5] lend themselves readily to such probabilistic modeling. In the area of underwater sound propagation, small random perturbations of the sound speed profile produce significant effects upon the acoustic signal if the propagation occurs over a long range [6].

In the analysis of such problems we seek statistical information about physically-relevant quantities, e.g. moments of the signal intensity, correlation functions, etc. A common feature of the problems cited is the fact that they all involve small random inhomogeneities and long propagation paths. Another common trait is the fact that they are linear stochastic boundary value problems. Invariably, an excitation is prescribed at one position while a termination or radiation condition is specified at another.

This paper will consider such wave propagation problems in the order of increasing difficulty. The major part of the discussion will be devoted to the source-excited one dimensional slab problem; this problem can be analyzed in a reasonably complete manner. The more difficult problem of propagation in randomly perturbed metallic waveguides will be considered next. For this problem, several approximations must be made before useful information can be extracted. These approximations will be assessed qualitatively; this discussion will serve as background for the more difficult problem of randomly perturbed dielectric waveguides. The problem of long range underwater sound propagation is essentially the same as propagation in a dielectric waveguide; it will be discussed in the second paper [1].

In modeling the random perturbations, e.g. waveguide misalignments or soundspeed fluctuations, we wish to be reasonably general in the characterization since detailed statisical information is usually not available. It is usually physically plausible to assume that the perturbing process is almost surely bounded and wide sense stationary with zero mean. Another assumption which is reasonable from a physical point of view is that the perturbations should possess an asymptotic independence property, i.e. if we sample the perturbing process at two points, the ensuing random variables should tend to become statistically independent as the distance between the sampling points tends to infinity. A precise formulation of this property, known as the strong mixing property, is well documented in the literature (cf. [7] or Appendix A) and proves essential to the mathematical analysis.

We shall therefore consider these wave propagation problems in the asymptotic limit of small random perturbations and long propagation paths, where the underlying perturbing process is assumed to possess the strong mixing property. We seek asymptotic statistical information about the propagating wave field, i.e. the random solution of the stochastic boundary value problem. The relevant asymptotic limit was developed on a heuristic basis by Stratonovich [8], made mathematically precise by Khasminskii [9] and extended to an operator-theoretic framework by Papanicolaou and Varadhan [7].

In addition to studing the stochastic propagation problems per se, we are also interested in determining their connection with radiative transport theory. This latter theory, which describes the propagation and scattering of incoherent radiation intensities, is formulated phenomenologically on the basis of energy conservation and an assumed scattering mechanism [10]. This theory has met with considerable success in modeling physical phenomena and is considerably simpler to work with than the stochastic wave formulation. Therefore, it is of interest to ascertain the regime in which the stochastic theory leads to a transport theoretic approximation and also to determine the precise structure of the resulting transport equation (cf. [11], [12]).

In the next section, we shall carry out this program fully for the source-excited one dimensional slab problem, i.e. solve the stochastic wave problem and contrast the results both with the predictions of radiative transport theory and the results of numerical simulations. In Appendix A, the basic limit theorem upon which the analysis rests is presented. However, to convey the basic idea underlying the asymptotic limit we shall digress and briefly consider an elementary example.

EXAMPLE. Consider, on the one hand, the simplest of Ito stochastic differential equations

$$dX(t) = dW(t), \qquad X(0) = x_0, \tag{1}$$

where $W(t)$ denotes a standard Wiener process, the mathematical model of Brownian motion. As is customary, we suppress the dependence upon the variable indexing the probability space. The solution

$$X(t) = x_0 + W(t) \tag{2}$$

is, at fixed t, a normally distributed random variable with mean x_0 and variance t. More generally, the solution process is a Gauss-Markov diffusion process. Expectations of suitable functions (e.g. moments) of the solution process, $E\{f(X(t))\}$, can be computed by solving

$$\frac{\partial}{\partial t}u(t,x) = \frac{1}{2}\frac{\partial^2}{\partial x^2}u(t,x), \quad u(0,x) = f(x) \qquad (3)$$

and evaluating the solution at $x = x_0$. Thus, the behavior of the solution process at the probabilistic level is quite nice. Note, however, that at the level of sample functions or individual realizations, the solution paths possess the pathological structure of the Wiener paths [14].

To develop an asymptotic analog of (1), we shall begin by defining the piecewise constant stochastic process

$$a(t) \equiv a_j, \quad j - 1 \leqslant t < j, \quad j = 1,2,\cdots, \qquad (4)$$

where the random variables $\{a_j\}_{j=1}^{\infty}$ are assumed to constitute a set of independent, identically distributed, a.s. bounded random variables having mean 0 and variance 1. The initial value problem that we consider is

$$\frac{d}{dt}X(t) = \varepsilon a(t), \quad X(0) = x_0, \qquad (5)$$

where ε is a small real parameter. At $t = N$, the solution of (5) is

$$X(N) = x_0 + \varepsilon \sum_{j=1}^{N} a_j. \qquad (6)$$

Note that (6), together with the Central Limit Theorem, suggests the appropriate asymptotic limit. If we define $\tau \equiv \varepsilon^2 t$, then, at $t = N$, $\varepsilon = \sqrt{\tau/N}$ and

$$X(\tau/\varepsilon^2) = x_0 + \sqrt{\tau}\left[\frac{\sum_{j=1}^{N} a_j}{\sqrt{N}}\right]. \qquad (7)$$

Thus, if we let $\varepsilon \to 0$ while keeping τ fixed, then $N \to \infty$ and the right side of (7) converges in distribution to a normally distributed random variable having mean x_0 and variance τ.

In this simple example (cf. [15], p. 452) and in the general asymptotic theory as well, we actually obtain more, i.e. weak convergence on finite τ intervals to a Markov diffusion process. In [16], the convergence of moments is established for the general asymptotic theory. We are again able to exploit the theory of partial differential equations to deduce information about the (asymptotic) behavior of expectations of functions of the solution process. Note also that while the asymptotic limit exists at the probabilistic level, it does not exist at the level of paths. In the τ variable, (5) becomes

$$\frac{d}{d\tau} X(\tau/\varepsilon^2) = \frac{1}{\varepsilon} a(\tau/\varepsilon^2), \quad X(0) = x_0. \qquad (8)$$

Thus, the asymptotic limit mimics, at the level of paths, the pathological structure of the Wiener paths.

2. One-Dimensional Slab of Random Medium.

The one-dimensional problem that we shall consider is illustrated in Figure 1.
A source is located within a slab of length ℓ whose index of refraction is randomly perturbed; beyond the slab extremities, energy radiates outward to infinity.

This problem was initially studied by A. Schuster [17] in 1905. Since then, it has been studied by many investigators, including Gertsenshtein and Vasil'ev [18], [19], Gazaryan [20], Lang [21], Marcuse [22] and Morrison, Papanicolaou and Keller [23]. The present discussion, based on work by the authors [24], [25] draws particularly from these cited works. We assume a monochromatic $e^{-i2\pi ft}$ time dependence and let $u(x)$ denote the complex-valued scalar wave field of interest. The problem to be studied is

$$\frac{d^2}{dx^2}u(x) + k^2[1 + \varepsilon\mu(x)]u(x) = i2k\delta(x - y), \quad 0 \leqslant x, \quad y \leqslant \ell;$$

$$u(x) = T_- e^{-ikx}, \quad x \leqslant 0; \quad u(x) = T_+ e^{ikx}, \quad x \geqslant \ell;$$

$$u(x) \text{ and } \frac{d}{dx}u(x) \text{ continuous across } x = 0 \text{ and } \ell. \tag{9}$$

The differential equation is an inhomogeneous reduced wave equation-with $k = 2\pi/\lambda$ the free space wavenumber. The source strength has been chosen so that when $y = 0$, i.e. the source is at the left end of the slab, the problem reduces to that of a random slab illuminated from the left by a plane wave of unit

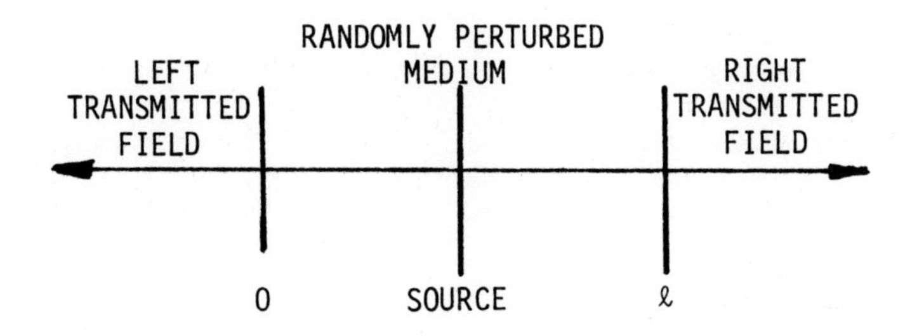

FIGURE 1: SLAB CONFIGURATION

amplitude. Again, ε is a small real parameter and the stochastic process $\mu(x)$ used to model the refractive index perturbations is assumed to be a.s. bounded, zero mean, wide sense stationary and strongly mixing. Thus, we have

$$E\{\mu(x)\} = 0, \qquad E\{\mu(x+z)\mu(x)\} = R(z). \qquad (10)$$

In order to exploit the asymptotic theory we shall reformulate the problem, removing the rapid phase variation. We introduce a travelling wave formulation, defining slowly-varying complex envelopes $A(x,y,\ell)$ and $B(x,y,\ell)$ via the variation of parameters prescription

$$u(x) = A(x)e^{ikx} + B(x)e^{-ikx},$$

$$\frac{d}{dx}u(x) = ik[A(x)e^{ikx} - B(x)e^{-ikx}]. \qquad (11)$$

In the absence of the random perturbations, i.e. if $\varepsilon = 0$, the quantities A and B would be constants. Because of the perturbations, they are slowly-varying functions of position. In terms of these new dependent variables, problem (9) becomes

$$\frac{d}{dx}\begin{bmatrix} A(x) \\ B(x) \end{bmatrix} = \varepsilon m(x)\begin{bmatrix} A(x) \\ B(x) \end{bmatrix}, \quad m(x) \equiv \frac{ik\mu(x)}{2}\begin{bmatrix} 1 & e^{-i2kx} \\ -e^{i2kx} & -1 \end{bmatrix},$$

$$\left.\begin{aligned} A(y+0,y,\ell) - A(y-0,y,\ell) &= e^{-iky} \\ B(y+0,y,\ell) - B(y-0,y,\ell) &= -e^{iky} \end{aligned}\right\} \quad \text{(jump conditions across source),}$$

$$A(0,y,\ell) = B(\ell,y,\ell) = 0 \qquad \text{(boundary conditions at the slab extremities).} \tag{12}$$

For this problem, the quantities of interest are the mean values and fluctuations of observables, such as the total intensity of radiation and the power fluxes. We define the total intensity by

$$J(x,y,\ell) = |A(x,y,\ell)|^2 + |B(x,y,\ell)|^2 = \tfrac{1}{2}\left(|u|^2 + \frac{1}{k^2}\left|\frac{du}{dx}\right|^2\right). \tag{13}$$

This quantity depends on the observation point, source point and slab length. Note that when $\varepsilon = 0$, then $u(x) = e^{ik|x-y|}$, $|A|^2 - |B|^2 = \mathrm{sgn}(x-y)$ and $|A|^2 + |B|^2 = 1$. The right and left power fluxes are given by

$$\begin{aligned} |T_+(y,\ell)|^2 &= |A(x,y,\ell)|^2 - |B(x,y,\ell)|^2 = J(\ell,y,\ell), \quad x > y, \\ |T_-(y,\ell)|^2 &= -|A(x,y,\ell)|^2 + |B(x,y,\ell)|^2 = J(0,y,\ell), \quad x < y. \end{aligned} \tag{14}$$

As the notation suggests, the power flux becomes independent of the observer position (once we specify the ordering of source and observer points), which is to be expected since the slab itself neither creates nor dissipates energy. This fact, together with boundary conditions (12), enables one to compute the power fluxes as special evaluations of the total intensity. We shall also compute correlations of the complex envelope functions at different observer points, e.g. $E\{A(x_1,y,\ell)\overline{A}(x_2,y,\ell)\}$, $E\{A(x_1,y,\ell)\overline{B}(x_2,y,\ell)\}$, etc.

DERIVATION OF RESULTS. Consider the fundamental matrix or propagator solution corresponding to differential equation (12), i.e.

$$\frac{d}{dx}Y(x,x') = \varepsilon m(x)Y(x,x'), \quad Y(x',x') = I, \quad x \geq x'. \quad (15)$$

Since $\operatorname{Tr} m(x) = 0$, $\det Y(x,x') = 1$. More specifically, the structure of m implies that Y has the form

$$Y = \begin{pmatrix} a & b \\ \bar{b} & \bar{a} \end{pmatrix}, \quad \det Y = |a|^2 - |b|^2 = 1. \quad (16)$$

Thus, Y is a stochastic process with values in the group $SU(1,1)$.

For the sake of definiteness, we shall assume that the observation point lies to the right of the source point, $0 \leq y \leq x \leq \ell$; the other case is basically similar (cf. [25]). The slab is thus divided into three regions, which we shall index numerically. We decompose the fundamental matrix $Y(\ell,0)$ into the product

$$Y(\ell,0) = Y_3(\ell,x)Y_2(x,y)Y_1(y,0), \quad (17)$$

where the matrices Y_3, Y_2, Y_1, when viewed as functions of their first argument, satisfy (15). Using the notation

$$Y_j = \begin{pmatrix} a_j & b_j \\ \bar{b}_j & \bar{a}_j \end{pmatrix}, \quad |a_j|^2 - |b_j|^2 = 1, \quad j = 1,2,3, \quad (18)$$

we obtain the following expressions for the complex envelopes $A(x,y,\ell)$ and $B(x,y,\ell)$

$$A = \frac{\bar{a}_3[\bar{a}_1 e^{-iky} + b_1 e^{iky}]}{b_1(\bar{b}_2\bar{a}_3 + a_2\bar{b}_3) + \bar{a}_1(b_2\bar{b}_3 + \bar{a}_2\bar{a}_3)},$$

$$B = \frac{-b_3[a_1 e^{-iky} + b_1 e^{iky}]}{b_1(\bar{b}_2\bar{a}_3 + a_2\bar{b}_3) + \bar{a}_1(b_2\bar{b}_3 + \bar{a}_2\bar{a}_3)},$$

$$0 \leqslant y \leqslant x \leqslant \ell. \qquad (19)$$

Thus, the quantities A and B are functions of three non-overlapping increments of the matrix-valued process Y. We shall also use the notation

$$\begin{pmatrix} a_{32} & b_{32} \\ \bar{b}_{32} & \bar{a}_{32} \end{pmatrix} = \begin{pmatrix} a_3 & b_3 \\ \bar{b}_3 & \bar{a}_3 \end{pmatrix} \begin{pmatrix} a_2 & b_2 \\ \bar{b}_2 & \bar{a}_2 \end{pmatrix}. \qquad (20)$$

It proves convenient to introduce the polar coordinates ϕ, ψ, θ by defining

$$a_j \equiv e^{i(\phi_j + \psi_j)/2} \cosh(\theta_{j/2}),$$

$$b_j \equiv e^{i(\phi_j - \psi_j)/2} \sinh(\theta_{j/2}), \quad j = 1,2,3, \quad (21)$$

with similar representations for a_{32}, b_{32} in terms of ϕ_{32}, ψ_{32}, θ_{32}. Here $0 \leqslant \phi_j < 2\pi$, $0 \leqslant \psi_j < 4\pi$, $0 \leqslant \theta_j < \infty$; when convenient, however, we may assume that $\psi_j \in [0,2\pi)$

and $\phi_j \in [0,4\pi)$. In terms of this angular parameterization, the radiation intensity $J(x,y,\ell)$ becomes

$$J = \frac{2 \cosh \theta_3[\cosh \theta_1 + \cos(2ky + \phi_1)\sinh \theta_1]}{1 + \cosh \theta_1 \cosh \theta_{32} + \cos(\phi_1 + \psi_{32})\sinh \theta_1 \sinh \theta_{32}},$$

$$0 \leqslant y \leqslant x \leqslant \ell. \qquad (22)$$

Note that

$$J \leqslant \cosh \theta_3(\cosh \theta_1 + \sinh \theta_1) = (|a_3|^2 + |b_3|^2)(|a_1| + |b_1|)^2.$$

Therefore, the intensity is a polynomially-bounded function of the non-overlapping increments of Y. It is precisely this class of function to which the limit theorem stated in Appendix A can be applied.

We define the following scaled variables

$\tau \equiv \varepsilon^2\ell$, scaled slab length,

$\xi \equiv \varepsilon^2 y-\tau/2$, scaled distance of source from slab center,

$\eta \equiv \varepsilon^2 x-\tau/2$, scaled distance of observer from slab center.

(23)

In terms of these new variables, the problem configuration becomes that of Figure 2.

We now consider $E\{J(\varepsilon^{-2}(\eta + \tau/2), \varepsilon^{-2}(\xi + \tau/2), \varepsilon^{-2}\tau)\}$ in the asymptotic limit, where $\varepsilon \to 0$ and ξ, η, τ remain fixed (with $-\frac{\tau}{2} \leqslant \xi \leqslant \eta \leqslant \frac{\tau}{2} < \infty$ for the particular configur-

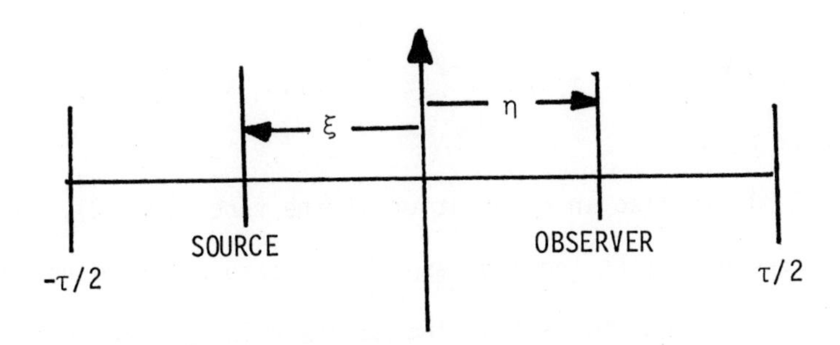

FIGURE 2: CONFIGURATION IN TERMS OF NEW VARIABLES

ation being studied). Recall that J is a certain polynomially-bounded function of three non-overlapping increments of the matrix-valued process Y; the explicit form for the intensity is given in (22). An application of the limit theorem cited in Appendix A leads to the conclusion that the asymptotic limit exists; i.e.

$$\lim_{\substack{\varepsilon\to 0\\ \xi,\eta,\tau \text{ fixed}}} E\{J(\varepsilon^{-2}(\eta + \tau/2),\varepsilon^{-2}(\xi + \tau/2),\varepsilon^{-2}\tau)\} \equiv MJ(\tau,\xi,\eta), \quad (24)$$

where MJ is also an expectation of the function (22). However, the variables and indices in this case refer to non-overlapping increments of a limiting matrix-valued Markov diffusion process, which we shall call Y_0. This limiting process (cf. [24]), moreover, has stationary independent matrix increments; i.e., if $0 \leqslant \sigma_1 \leqslant \sigma_2 \leqslant \cdots \leqslant \sigma_n$, then the random matrices

$$Y_0(\sigma_1), Y_0(\sigma_2)Y_0^{-1}(\sigma_1), \cdots, Y_0(\sigma_n)Y_0^{-1}(\sigma_{n-1})$$

are statistically independent and their distribution depends only upon $\sigma_1, \sigma_2-\sigma_1, \cdots, \sigma_n-\sigma_{n-1}$. If we define the parameters

$$\alpha \equiv \frac{k^2}{2}\int_0^\infty R(s)\cos 2ks\, ds \geqslant 0,$$

$$\beta \equiv \frac{k^2}{2}\int_0^\infty R(s)\sin 2ks\, ds,$$

$$\gamma \equiv k^2 \int_0^\infty R(s)ds > 0, \quad (25)$$

then the infinitesimal generator associated with the limiting process Y_0 is

$$V = \alpha[\frac{\partial^2}{\partial\theta^2} + \coth\theta\frac{\partial}{\partial\theta} + (\coth\theta\frac{\partial}{\partial\phi} - \operatorname{csch}\theta\frac{\partial}{\partial\psi})^2] + \gamma\frac{\partial^2}{\partial\phi^2} - \beta\frac{\partial}{\partial\phi}. \quad (26)$$

Note that the parameters α and γ are proportional to power spectrum evaluations (at wave numbers $2k$ and 0 respectively) and thus are non-negative.

To compute the asymptotic behavior of the mean intensity, i.e. $MJ(\tau,\xi,\eta)$, we must compute the expected value of (22), where the indices refer to the stationary independent increments of the diffusion Markov process Y_0; this process, in turn, is characterized by the generator V given in (26) and the specification of $Y_0(0)$. The nice aspect of this problem is that results from special function theory can be used to obtain an explicit formula for MJ. Moreover, an asymptotic expression for the second moment of the radiation intensity can likewise be computed. We shall adopt the notation

$$\lim_{\varepsilon\to 0} E\{J^2(\varepsilon^{-2}(\eta + \tau/2),\varepsilon^{-2}(\xi + \tau/2),\varepsilon^{-2}\tau)\} \equiv KJ(\tau,\xi,\eta),$$

$$FJ(\tau,\xi,\eta) \equiv \left(KJ(\tau,\xi,\eta) - [MJ(\tau,\xi,\eta)]^2\right)^{1/2}. \quad (27)$$

The function FJ provides a measure of the intensity fluctuations within the slab in the asymptotic limit and enables us to assess the utility of MJ as an indicator of the actual behavior of the random field.

The derivation of the expressions for MJ and KJ (for both orderings of source and observations points) is presented in [24] and [25]; it will not be repeated here. However, in Appendix B, an analogous computation for a two-point correlation is outlined. The results obtained for the mean intensity and its second moment are

$$MJ(\tau,\xi,\eta) = e^{(3\alpha\tau/4 - \alpha|\xi-\eta|)}\int_{-\infty}^{\infty} e^{-t^2\alpha\tau} \frac{\pi \sinh \pi t}{t \cosh^2 \pi t}$$

$$\cdot[(t^2+1/4)\cos 2t\alpha(\xi+\eta) + (t^2-1/4)\cos 2t\alpha(\tau-|\xi-\eta|)$$

$$+ t \sin 2t\alpha(\tau-|\xi-\eta|)]dt, \qquad (28)$$

$$KJ(\tau,\xi,\eta) = \frac{e}{8}^{(15\alpha\tau/4-4\alpha|\eta-\xi|)}\int_{-\infty}^{\infty} e^{-t^2\alpha\tau} \frac{\pi\ t \sinh \pi t}{\cosh^2 \pi t}$$

$$\cdot\frac{(t^2+1/4)}{(t^2+1)^2}\left|\left[(t^2+\tfrac{3}{4})e^{-2\alpha(\tau-2\sigma\eta)} + (t^2+\tfrac{5}{4})\cos 2t\alpha(\tau-2\sigma\eta) + (t^2+\tfrac{3}{4})\right.\right.$$

$$\left.\left.\cdot\frac{\sin 2t\alpha(\tau-2\sigma\eta)}{t}\right]\right|\left[(t^2+\tfrac{1}{4})e^{-2\alpha(\tau+2\sigma\xi)} + 3(t^2+\tfrac{5}{4})\cos 2t\alpha(\tau+2\sigma\xi)\right.$$

$$\left. + (t^2+\tfrac{3}{4})\frac{\sin 2t\alpha(\tau+2\sigma\xi)}{t}\right]dt, \quad \sigma \equiv \mathrm{sgn}(\eta - \xi). \qquad (29)$$

Note that both MJ and KJ are actually functions of $\alpha\tau$, $\alpha\xi$ and $\alpha\eta$, where α is the power spectrum evaluation defined by (25). Observe also that the exponential factors multiplying the integrals in (28) and (29) will increase with increasing $\alpha\tau$ unless the source and observation points are, roughly speaking, at opposite ends of the slab.

The important special case of power transmission through

a slab illuminated on the left by a plane wave of unit amplitude can be studied by setting $\xi = -\tau/2$ and $\eta = \tau/2$ in (28) and (29). The mean power transmission coefficient becomes

$$MJ(\tau,-\tau/2,\tau/2) = e^{-\alpha\tau/4}\int_{-\infty}^{\infty} e^{-t^2\alpha\tau} \frac{\pi t \sinh \pi t}{\cosh^2 \pi t}\, dt \qquad (30)$$

and its second moment is

$$KJ(\tau,-\tau/2,\tau/2) = e^{-\alpha\tau/4}\int_{-\infty}^{\infty} e^{-t^2\alpha\tau} \frac{\pi t \sinh \pi t}{\cosh^2 \pi t}\left(t^2 + \frac{1}{4}\right)dt. \qquad (31)$$

In this case, since $\eta - \xi = \tau$, the moments are exponentially decreasing functions of $\alpha\tau$. The mean intensity within the slab corresponding to plane wave illumination from the left, i.e.

$$MJ(\tau,-\tau/2,\eta) = e^{\alpha\tau/4-\alpha\eta}\int_{-\infty}^{\infty} e^{-t^2\alpha\tau} \frac{\pi t \sinh \pi t}{\cosh^2 \pi t}\left[\cos t\alpha(\tau-2\eta) + \frac{\sin t\alpha(\tau-2\eta)}{2t}\right]dt, \quad -\frac{\tau}{2} \leqslant \eta \leqslant \frac{\tau}{2}, \qquad (32)$$

was initially derived by Gazaryan [20] and Lang [21] using somewhat different approaches. Gazaryan also made the interesting observation that the expression for MJ given in (32) can be viewed as a function of τ,η for $\tau \geqslant 0$ and $-\infty < \eta < \infty$ and on this domain MJ is the unique solution of the Cauchy problem

$$\frac{\partial}{\partial(\alpha\tau)}MJ = \frac{1}{4}\frac{\partial^2}{\partial(\alpha\eta)^2}MJ, \quad MJ(0,0,\eta) = 1 - \tanh \alpha\eta - \alpha\eta\, \operatorname{sech}^2\alpha\eta. \qquad (33)$$

Although this observation has proven useful [24], no real underlying significance has yet been attached to it.

We conclude this section by summarizing the asymptotic behavior of the two point correlations. For simplicity, we consider only the case $\xi = -\tau/2$, i.e. the case of plane wave illumination from the left.

Let η_1 and $\eta_2 > \eta_1$ denote the two scaled observation points, measured from the slab center. We define the following

$P^{\nu}_{mn} \equiv$ generalized Legendre function ([26], Chapt. VI);

$$\nu \equiv -\frac{1}{2} + it,$$

$$p(\eta_2-\eta_1,u) \equiv \frac{e^{-(\gamma+\alpha+i2\beta)(\eta_2-\eta_1)/4}}{8\pi^3} \int_0^\infty e^{-\alpha(\eta_2-\eta_1)(t^2+\frac{1}{4})}$$

$$\cdot P^{\nu}_{1/2,1/2}(u)t \coth \pi t \, dt,$$

(cf. (25) for definitions of α,β,γ). (34)

$$M_{\pm}(t,\frac{\tau}{2} - \eta_2) \equiv [\cos t\alpha(\tau-2\eta_2) + \frac{\sin t\alpha(\tau-2\eta_2)}{2t}] \pm e^{-\alpha(\frac{\tau}{2} - \eta_2)},$$

$$N(t,\frac{\tau}{2} - \eta_2) \equiv \frac{\Gamma(\nu)}{\Gamma(\nu+2)} \cdot \frac{(t^2 + \frac{1}{4})}{t} \sin t\alpha(\tau-2\eta_2).$$

Then, the asymptotic behavior of the two-point correlations can be shown to be

$$\lim_{\varepsilon\to 0} E\{A(\varepsilon^{-2}(\eta_2 + \tau/2),0,\varepsilon^{-2}\tau)\overline{A}(\varepsilon^{-2}(\eta_1 + \tau/2),0,\varepsilon^{-2}\tau)\}$$

$$= \frac{e^{\alpha\tau/4-\alpha(3\eta_2+\eta_1)/4}}{2} \int_{-\infty}^{\infty} e^{-t^2\alpha(\tau-\eta_2+\eta_1)} \frac{\pi t \sinh \pi t}{\cosh^2 \pi t}[M_+(t,\tfrac{\tau}{2} - \eta_2)$$

$$\cdot \int_1^{\infty} \sqrt{(u+1)/2}\, P_\nu(u)p(\eta_2 - \eta_1,u)du - N(t,\tfrac{\tau}{2} - \eta_2) \int_1^{\infty} \sqrt{(u-1)/2}$$

$$\cdot P_\nu^1(u)p(\eta_2 - \eta_1,u)du]dt. \qquad (35)$$

$$\lim_{\varepsilon\to 0} E\{\overline{B}(\varepsilon^{-2}(\eta_2 + \tau/2),0,\varepsilon^{-2}\tau)B(\varepsilon^{-2}(\eta_1 + \tau/2),0,\varepsilon^{-2}\tau)\}$$

$$= \frac{e^{\alpha\tau/4-\alpha(3\eta_2+\eta_1)/4}}{2} \int_{-\infty}^{\infty} e^{-t^2\alpha(\tau-\eta_2+\eta_1)} \frac{\pi t \sinh \pi t}{\cosh^2 \pi t}[M_-(t,\tfrac{\tau}{2} - \eta_2)$$

$$\cdot \int_1^{\infty} \sqrt{(u+1)/2}\, P_\nu(u)p(\eta_2 - \eta_1,u)du - N(t,\tfrac{\tau}{2} - \eta_2) \int_1^{\infty} \sqrt{(u-1)/2}$$

$$\cdot P_\nu^1(u)p(\eta_2 - \eta_1,u)du]dt. \qquad (36)$$

$$\lim_{\varepsilon\to 0} E\{A(\varepsilon^{-2}(\eta_2 + \tau/2),0,\varepsilon^{-2}\tau)\overline{B}(\varepsilon^{-2}(\eta_1 + \tau/2),0,\varepsilon^{-2}\tau)\}$$

$$= \lim_{\varepsilon\to 0} E\{\overline{B}(\varepsilon^{-2}(\eta_2 + \tau/2),0,\varepsilon^{-2}\tau)A(\varepsilon^{-2}(\eta_1 + \tau/2),0,\varepsilon^{-2}\tau)\} = 0. \qquad (37)$$

In equations (35), (36), P_ν and P_ν^1 denote the Legendre and associated Legendre functions $(P_\nu^1(u) = (u^2-1)^{1/2} \frac{d}{du} P_\nu(u))$, respectively. Recall that in the asymptotic limit

$-\frac{\tau}{2} \leq \eta_1 \leq \eta_2 \leq \frac{\tau}{2} < \infty$ are fixed. Appendix B outlines a representative computation of this type.

RADIATION TRANSPORT FORMULATION. A. Schuster, in 1905, studied the radiative transport analog of the one-dimensional slab problem [17]. We shall now formulate and solve this phenomenological version.

Consider a slab of length τ with a source located at ξ and an observation point at η; the problem configuration, therefore, is that of Figure 2. The source excites incoherent intensities $I^{\pm}(\tau,\xi,\eta)$ which propagate to the right (+) and to the left (-) within the slab. Inhomogeneities within the slab scatter the radiation; the assumed scattering mechanism is depicted in Figure 3. (For simplicity, we suppress the τ and ξ dependence.)

In traversing a distance $\Delta\eta$, therefore, the right propagating intensity I^+ become diminished by $\alpha\Delta\eta I^+$ due to backscattering but is simultaneously augumented by $\alpha\Delta\eta I^-$ due to the forward scattering of I^-. Conservation of energy considerations then dictate the relation for I^-. We are thus led to the problem

$$\frac{d}{d\eta} I^+(\tau,\xi,\eta) = \frac{d}{d\eta} I^-(\tau,\xi,\eta) = -\alpha[I^+(\tau,\xi,\eta) - I^-(\tau,\xi,\eta)],$$

$$I^{\pm}(\tau,\xi,\xi+0) - I^{\pm}(\tau,\xi,\xi-0) = \pm 1 \quad \text{(jump conditions across the source)},$$

$$I^+(\tau,\xi,-\tau/2) = I^-(\tau,\xi,\tau/2) = 0 \quad \text{(boundary conditions at the slab extremities)}. \quad (38)$$

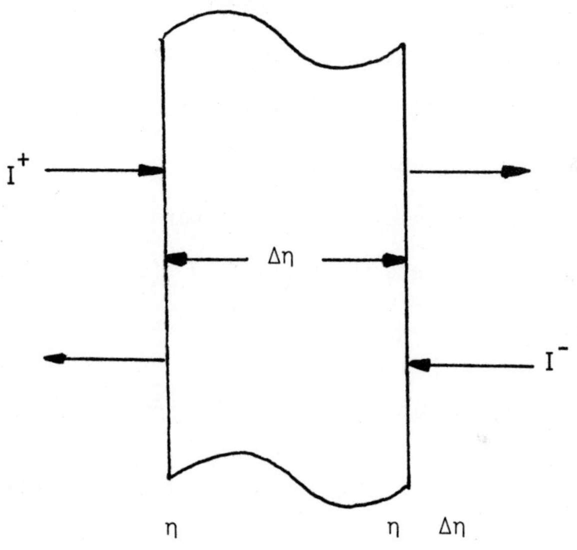

$$I^{+}(\eta + \Delta\eta) \cong I^{+}(\eta)(1 - \alpha\Delta\eta) + I^{-}(\eta)(\alpha\Delta\eta)$$

$$I^{-}(\eta) \cong I^{-}(\eta + \Delta\eta)(1 - \alpha\Delta\eta) + I^{+}(\eta)(\alpha\Delta\eta)$$

FIGURE 3: RADIATIVE TRANSPORT FORMULATION

Note that the scattering parameter has been tacitly chosen to be the constant α, defined in (25). This choice can be justified on an a posteriori basis as the one needed to put the stochastic wave solution and the transport result into some form of agreement. Observe also from (38) that the radiation flux, $I^+ - I^-$ if $\eta > \xi$ and $I^- - I^+$ if $\xi > \eta$, is independent of the observation point η.

The solution of (38) is straightforward. The total intensity, which we shall denote by $MJ_s(\tau,\xi,\eta)$ (s for Schuster), is found to be

$$MJ_s(\tau,\xi,\eta) \equiv I^+(\tau,\xi,\eta) + I^-(\tau,\xi,\eta) = \frac{[1+\alpha(\tau+2\sigma\xi)][1+\alpha(\tau-2\sigma\eta)]}{1+\alpha\tau},$$

$$\sigma \equiv \operatorname{sgn}(\eta - \xi). \qquad (39)$$

No analog of the second moment or fluctuations of the total intensity (i.e. KJ or FJ) exists for this model. A comparison of the expressions MJ and MJ_s reveals that the two expressions for the total intensity agree well in the case where $\alpha\tau$ is small. Specifically,

$$MJ(\tau,\xi,\eta) = MJ_s(\tau,\xi,\eta) + O((\alpha\tau)^3). \qquad (40)$$

By setting $\xi = -\tau/2$ and $\eta = \tau/2$, we obtain the power transmission coefficient, i.e.

$$MJ_s(\tau,-\tau/2,\tau/2) = \frac{1}{1+\alpha\tau}. \qquad (41)$$

Thus, the transport model leads to a transmission coefficient exhibiting algebraic decay with increasing $\alpha\tau$; the stochastic solution, on the other hand (cf. (30)), decays exponentially as a function of $\alpha\tau$.

Note that in the transport formulation, as in the stochastic wave problem, the right and left power fluxes (which we have observed are independent of η) can be determined as particular evaluations of MJ_s. Specifically,

$$|T_{+_s}(\xi,\tau)|^2 = I^+ - I^- = MJ_s(\tau,\xi,\tau/2) = \frac{1+\alpha(\tau+2\xi)}{1+\alpha\tau}, \quad \eta > \xi,$$

$$|T_{-_s}(\xi,\tau)|^2 = I^- - I^+ = MJ_s(\tau,\xi,-\tau/2) = \frac{1+\alpha(\tau-2\xi)}{1+\alpha\tau}, \quad \eta < \xi. \tag{42}$$

COMPARISON WITH NUMERICAL SIMULATIONS. We now compare the stochastic wave and radiative transport solutions with the results of numerical simulation. Similar numerical studies have been conducted by Marcuse [22], Morrison [27], and Frisch et. al. [28]; their results agree basically with those that we shall exhibit.

In the numerical study, expected values were approximated by computing an average over 100 realizations. Each realization was a slab consisting of 2000 unit-length sections. Within each realization, the index of refraction was assumed to be a two-state process, with states $\sqrt{1 \pm \varepsilon}$. The initial state (i.e. the value of the process at the left end of the slab) was chosen randomly; subsequent switching of states occurred randomly at intervals that were (approximately) exponentially distributed. The average number of unit-length sections between changes of the refractive index varied between 2.5 and 10; the wavenumber k was chosen to be 0.5. The parameter ε was not specified directly but rather was determined by the other variables. Typically, ε fell within the range $0.1 \leqslant \varepsilon \leqslant 0.3$.

The results that are present in Figures 4-11 pertain to the case $\alpha\tau = 3.0$. This is a representative value; results for other values of $\alpha\tau$ are shown in [25]. The variation of intensity and intensity fluctuations with observer position are presented for source locations at $\xi = -\tau/2$ (left end of slab), $-\tau/4$ and 0 (slab center). Power transmission coefficient and cross-correlation results are also shown for the case $\xi = -\tau/2$.

As one reviews the numerical results, the following points should be noted.

i) Although only the results for $\alpha\tau = 3.0$ are presented, the agreement between stochastic and transport theoretic results tended to deteriorate with increasing $\alpha\tau$.

ii) As the figures show, both the mean intensity and its fluctuations (i.e. MJ and FJ) have peak values that increase as the source is moved from the end of the slab to the center. The large values of intensity, which represents large values of stored nonpropagating energy, is a manifestation of what Frisch

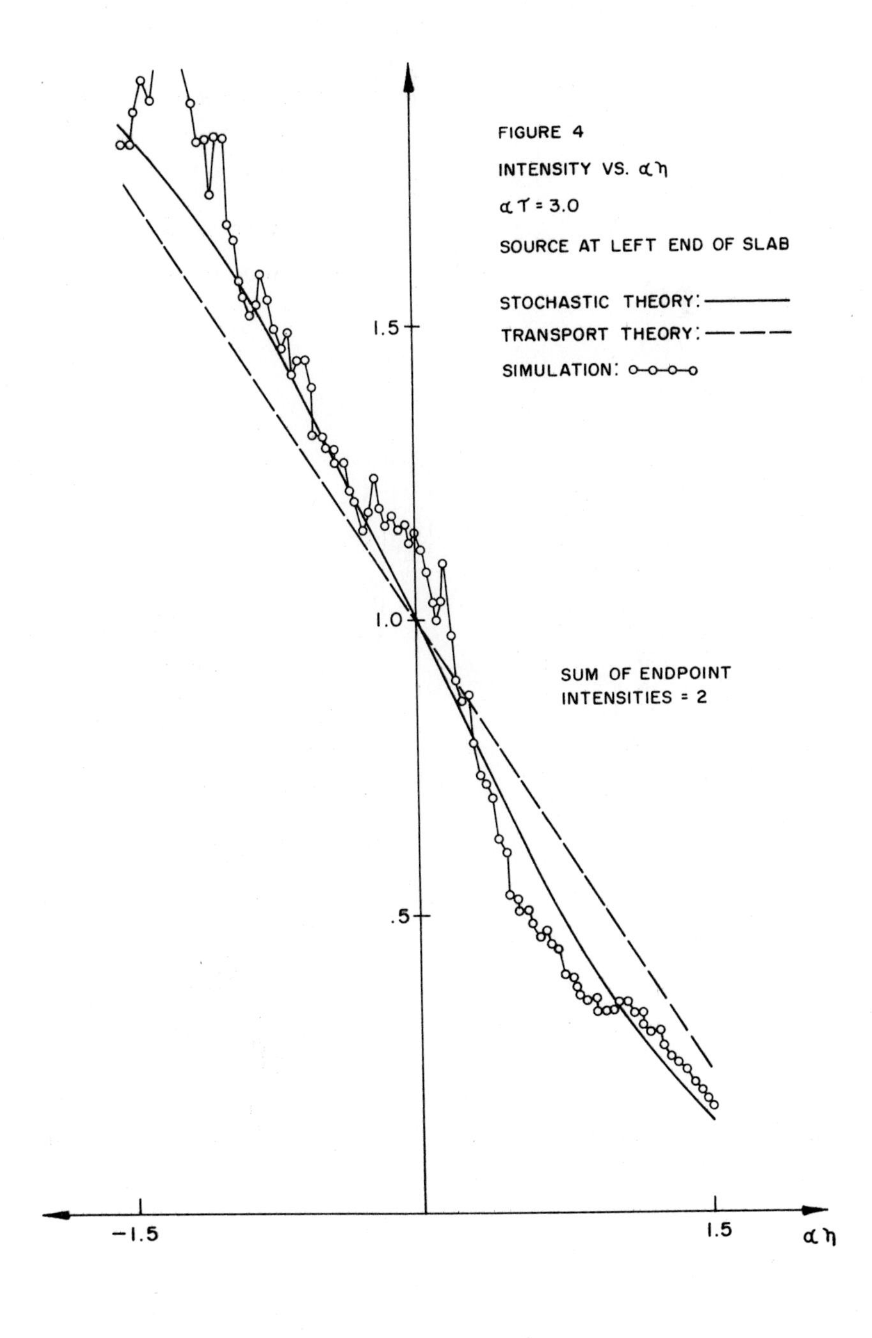
FIGURE 4
INTENSITY VS. $\alpha\eta$
$\alpha\tau = 3.0$
SOURCE AT LEFT END OF SLAB
STOCHASTIC THEORY:
TRANSPORT THEORY:
SIMULATION:
SUM OF ENDPOINT INTENSITIES = 2
1.5
1.0
.5
−1.5
1.5
$\alpha\eta$

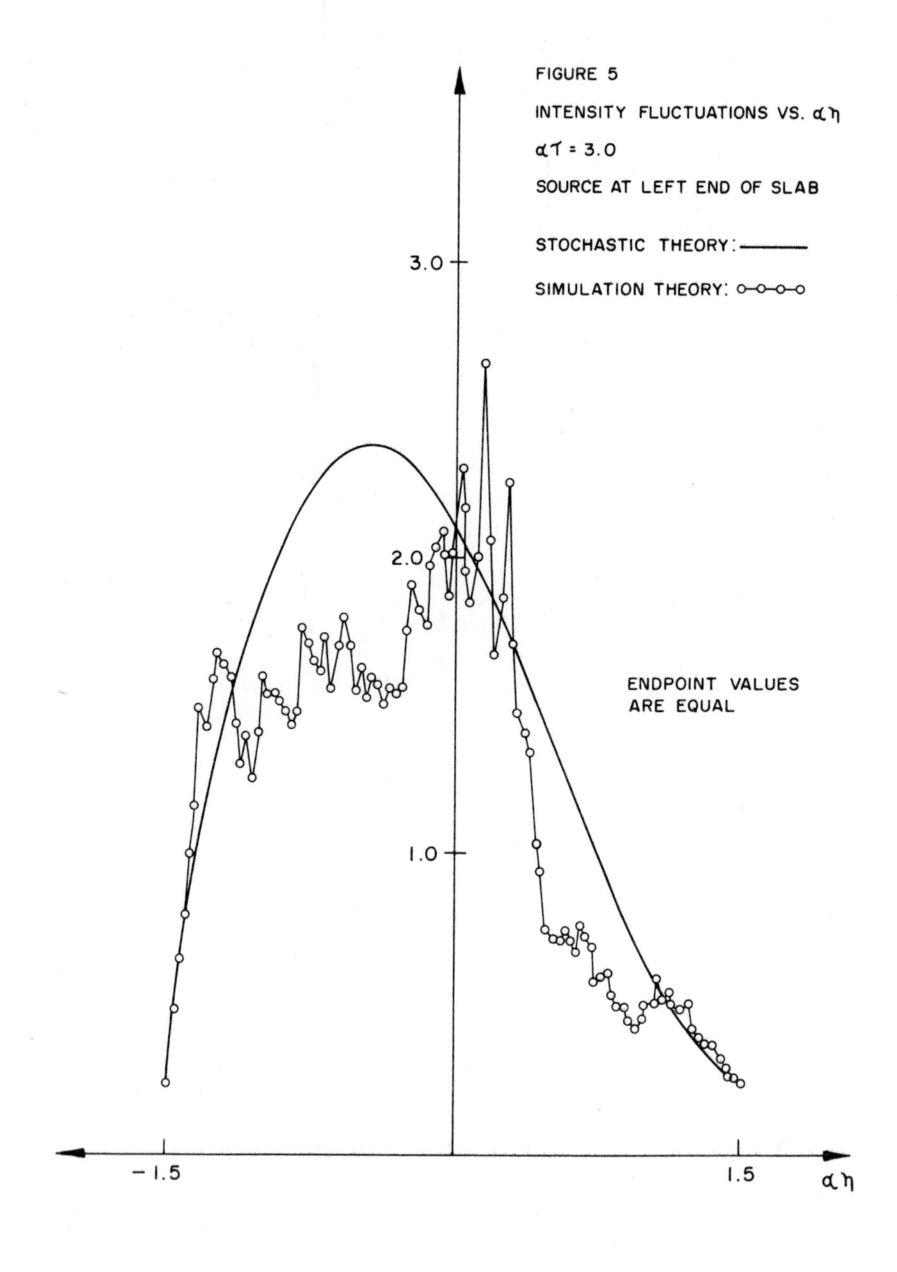
FIGURE 5
INTENSITY FLUCTUATIONS VS. $\alpha\eta$
$\alpha\tau = 3.0$
SOURCE AT LEFT END OF SLAB
STOCHASTIC THEORY:
SIMULATION THEORY:
3.0
2.0
1.0
ENDPOINT VALUES
ARE EQUAL
− 1.5
1.5
$\alpha\eta$

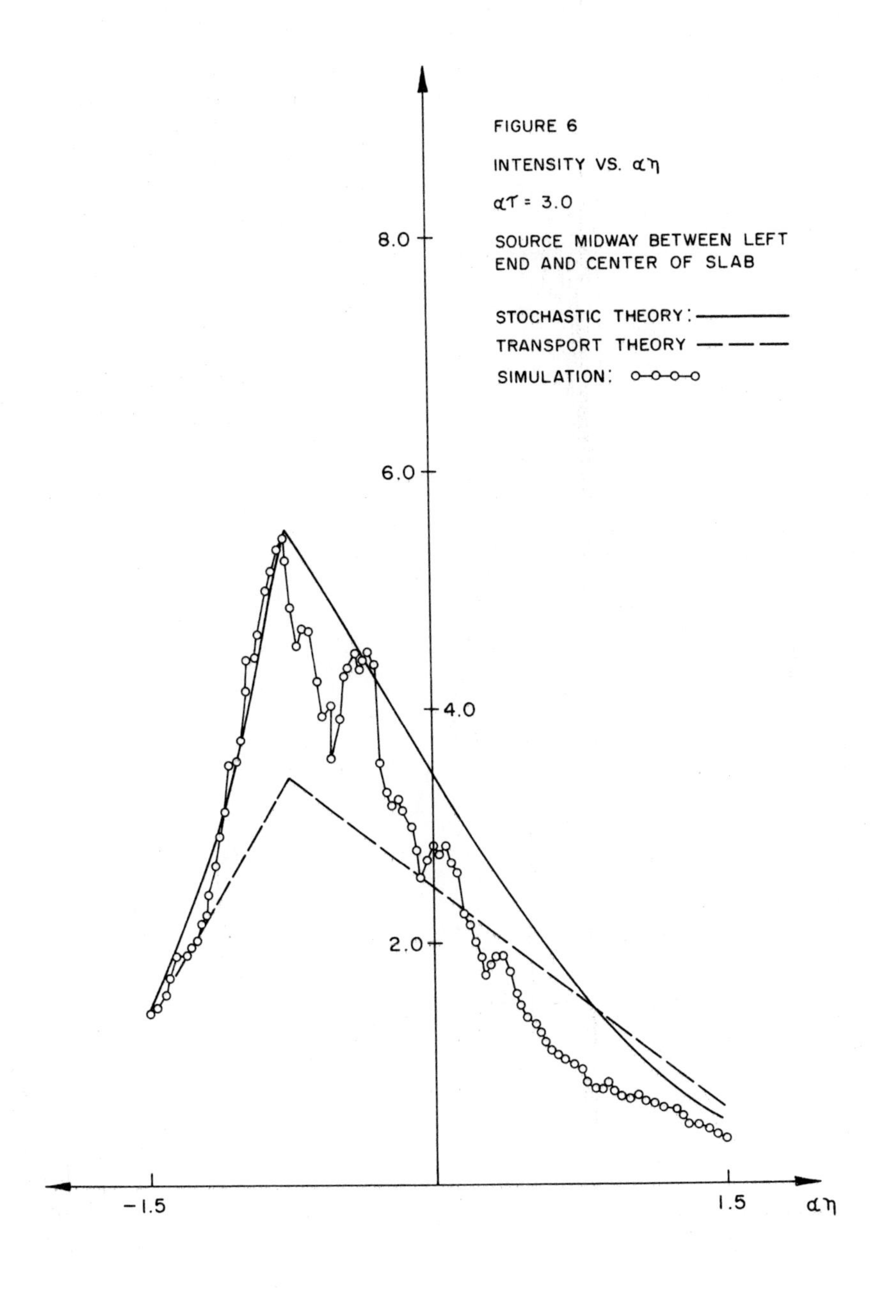
FIGURE 6
INTENSITY VS. $\alpha\eta$
αT = 3.0
SOURCE MIDWAY BETWEEN LEFT END AND CENTER OF SLAB
STOCHASTIC THEORY:
TRANSPORT THEORY
SIMULATION:
8.0
6.0
4.0
2.0
−1.5
1.5
$\alpha\eta$

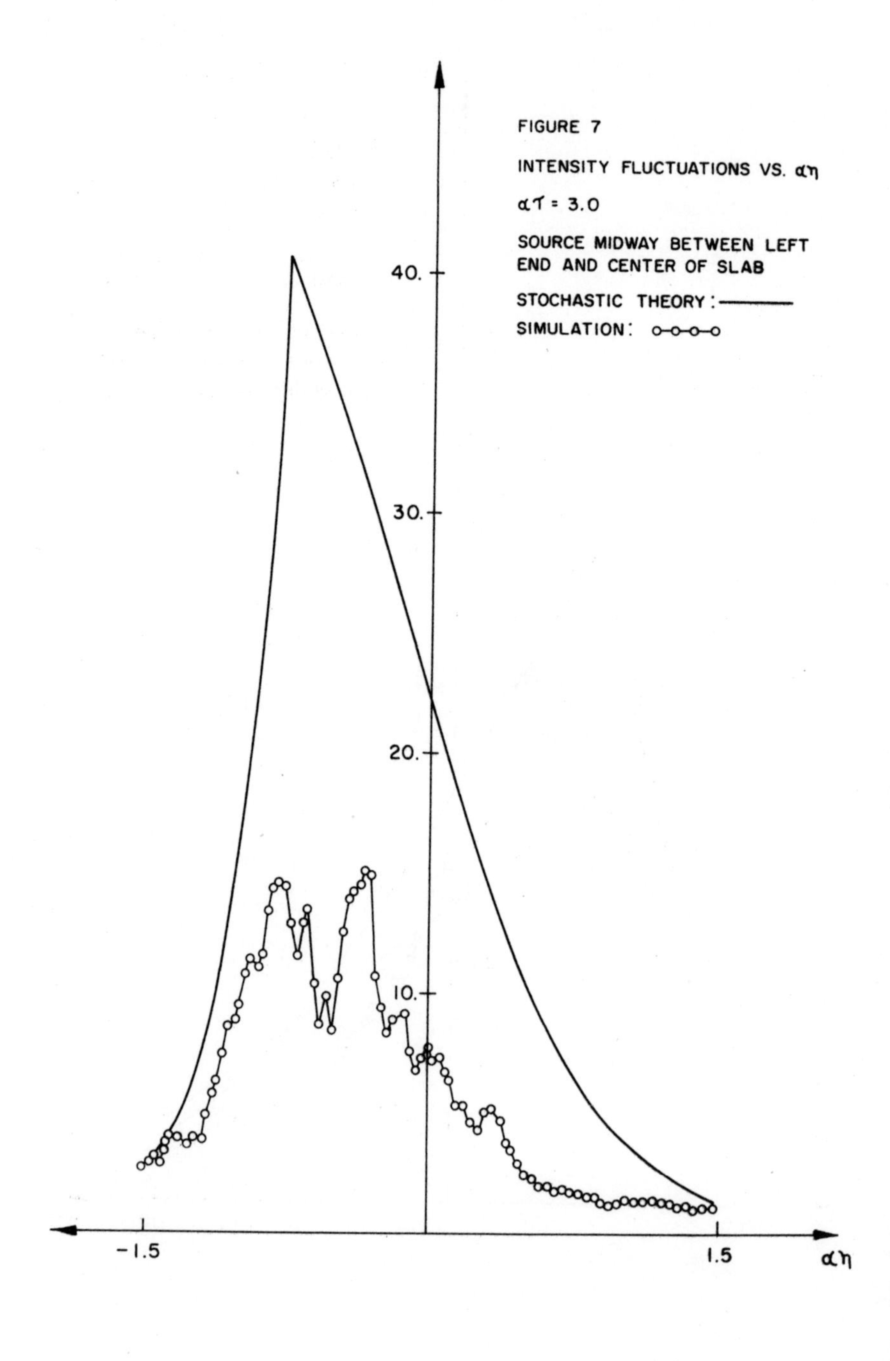
FIGURE 7
INTENSITY FLUCTUATIONS VS. αη
ατ = 3.0
SOURCE MIDWAY BETWEEN LEFT END AND CENTER OF SLAB
STOCHASTIC THEORY :
SIMULATION :
40.
30.
20.
10.
−1.5
1.5
αη

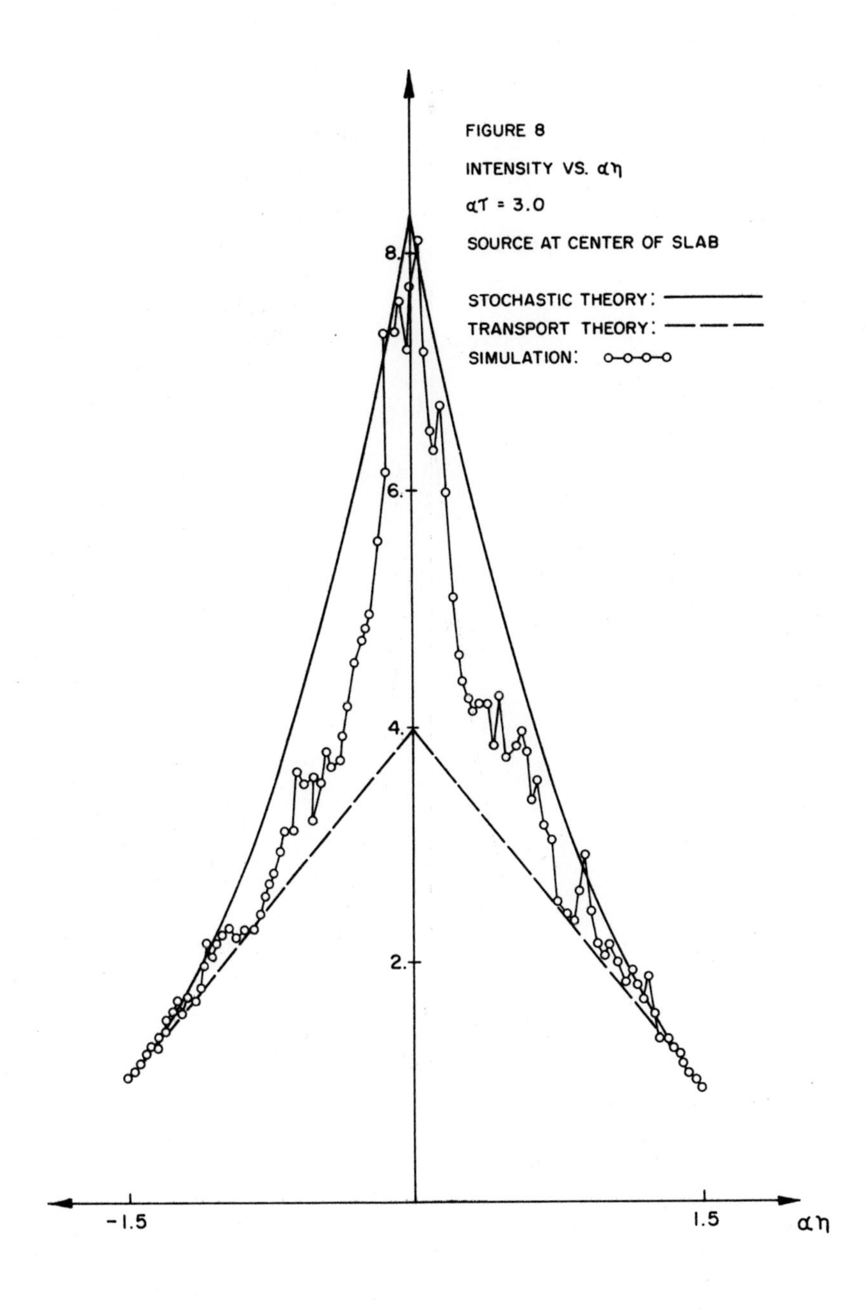
FIGURE 8
INTENSITY VS. αη
ατ = 3.0
SOURCE AT CENTER OF SLAB
STOCHASTIC THEORY:
TRANSPORT THEORY:
SIMULATION:
8.
6.
4.
2.
-1.5
1.5
αη

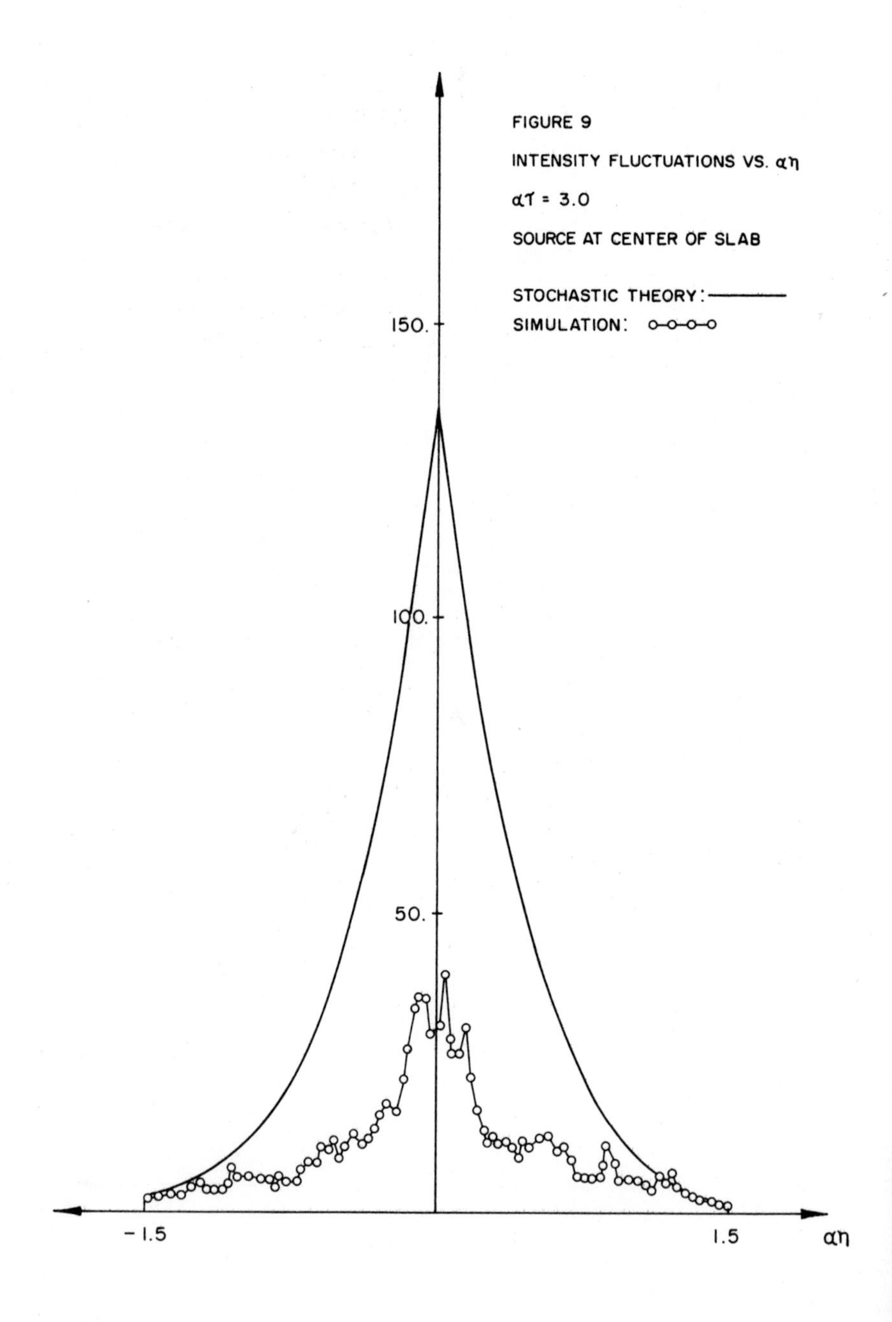
FIGURE 9
INTENSITY FLUCTUATIONS VS. αη
αT = 3.0
SOURCE AT CENTER OF SLAB
STOCHASTIC THEORY:
SIMULATION:
150.
100.
50.
− 1.5
1.5
αη

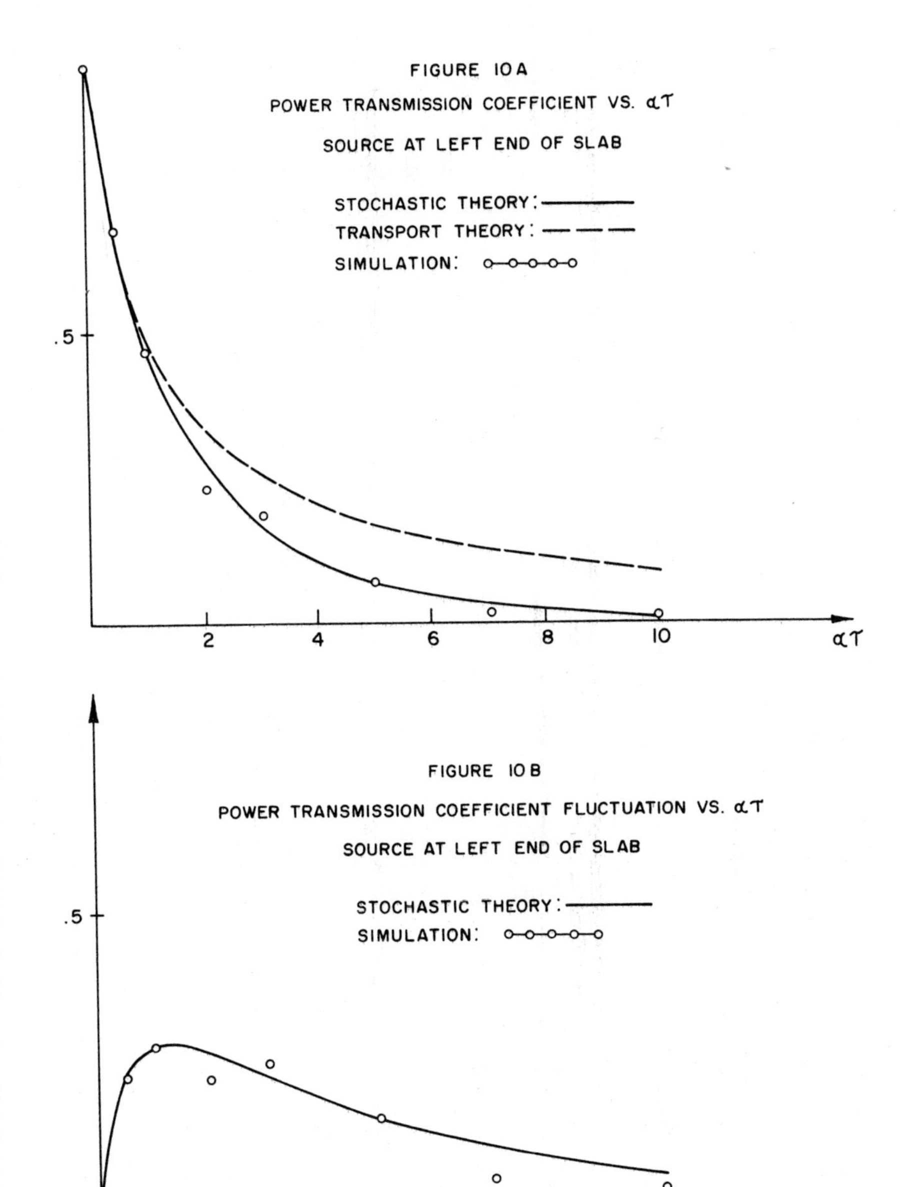
FIGURE 10A
POWER TRANSMISSION COEFFICIENT VS. ατ
SOURCE AT LEFT END OF SLAB
STOCHASTIC THEORY:
TRANSPORT THEORY:
SIMULATION:
.5
2
4
6
8
10
ατ
FIGURE 10B
POWER TRANSMISSION COEFFICIENT FLUCTUATION VS. ατ
SOURCE AT LEFT END OF SLAB
STOCHASTIC THEORY:
SIMULATION:
.5
2
4
6
8
10
ατ

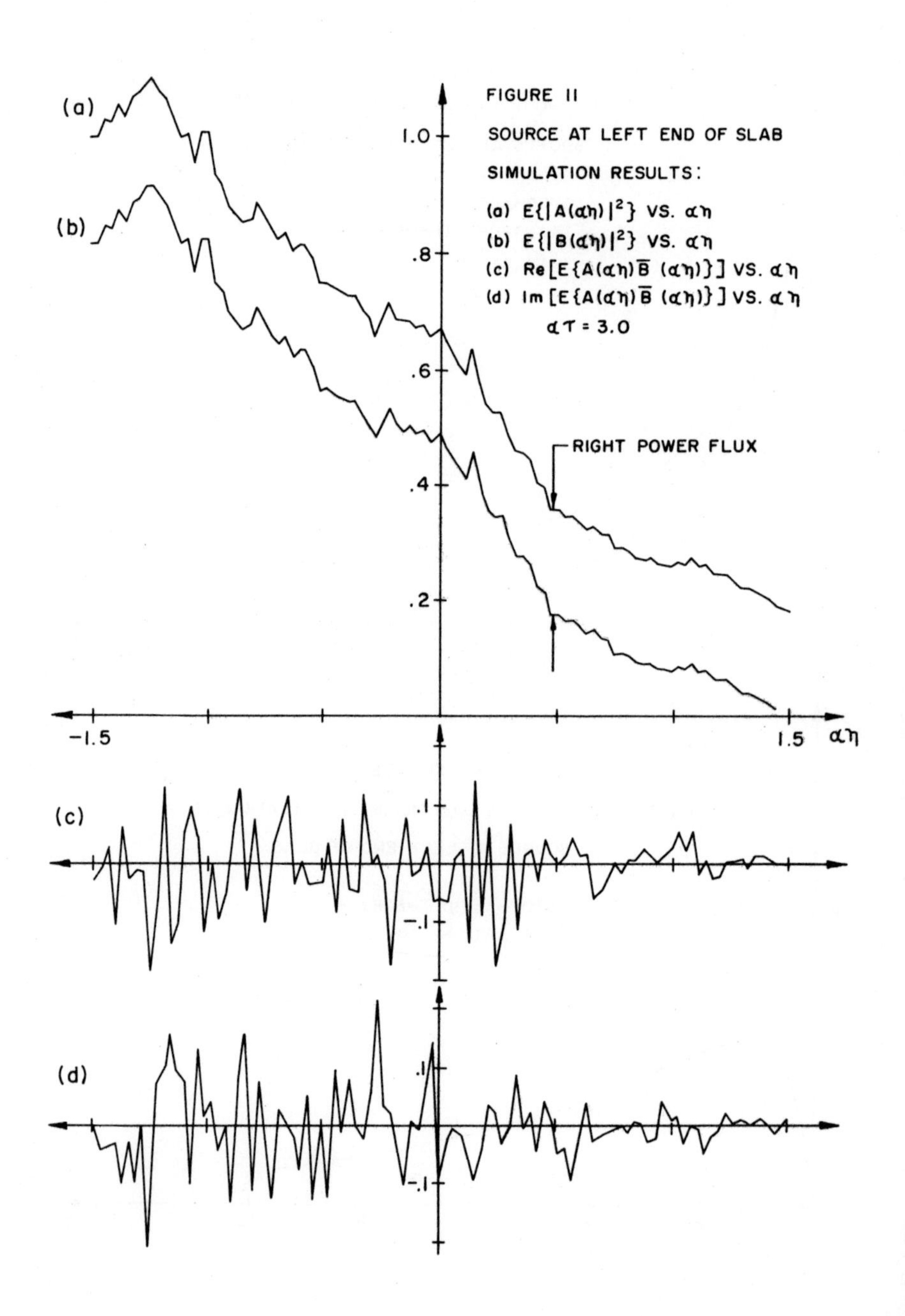
FIGURE II
SOURCE AT LEFT END OF SLAB
SIMULATION RESULTS:
(a) $E\{|A(\alpha\eta)|^2\}$ VS. $\alpha\eta$
(b) $E\{|B(\alpha\eta)|^2\}$ VS. $\alpha\eta$
(c) $Re[E\{A(\alpha\eta)\bar{B}(\alpha\eta)\}]$ VS. $\alpha\eta$
(d) $Im[E\{A(\alpha\eta)\bar{B}(\alpha\eta)\}]$ VS. $\alpha\eta$
$\alpha\tau = 3.0$
RIGHT POWER FLUX
(a)
(b)
(c)
(d)
1.0
.8
.6
.4
.2
−1.5
1.5
$\alpha\eta$
.1
−.1

et. al. [28] have termed "stochastic resonances."

iii) Although the mean intensity resulting from the stochastic theory agrees better with the numerical data than its transport theoretic counterpart, the agreement is somewhat illusory. Note that the intensity fluctuations FJ are much greater than the difference between MJ and MJ_s; in fact the fluctuations are comparable to MJ itself. Thus, it appears that the *mean intensity itself does not provide an adequate indicator of the field behavior within the slab.*

iv) The power transmission coefficient data in Figure 10 clearly distinguishes between the two theories. In this case, both the mean transmission coefficient and its fluctuations (cf. (27), (30), (31)) decrease exponentially with increasing $\alpha\tau$ in the stochastic theory while the transport theoretic result (41) predicts algebraic roll-off. As Figure 10 indicates, for $\alpha\tau \cong 10$ the spread between the two mean transmission coefficients is greater than the fluctuations associated with the stochastic theory. Thus, in this regime, we anticipate that the numerical data should clearly favor one of the two predictions; as Figure 10 shows, the numerical results are in close agreement with the predictions of stochastic theory.

v) The agreement between stochastic theory and the numerical simulations tended to deteriorate as the peak intensity and its fluctuations were increased, either by fixing the relative source position and increasing $\alpha\tau$ or by fixing $\alpha\tau$ and moving the source into the slab. This degradation in the agreement is believed to arise from two causes, the inherent difficulty in simulating strong resonances and the fact that for fixed ε, the error estimates for $|E\{J\} - MJ|$ and $|E\{J^2\} - KJ|$ themselves increase under these circumstances.

vi) In Figure 4 note that the sum of the total intensity at the left and right ends of the slab is 2. This stems from the fact that for this configuration (corresponding to a unit strength plane wave illuminating the slab from the left) the sum of the two intensities equals 1 + power reflection coefficient + power transmission coefficient. A consequence of this fact, in turn, is the fact that the endpoint fluctuations in Figure 5 are equal.

vii) Figure 11 presents simulation results for $E\{|A|^2\}$, $E\{|B|^2\}$ and $E\{A\bar{B}\}$. Note that since the source is at the left end of the slab, $E\{|A|^2\} - E\{|B|^2\}$ represents the (constant) right power flux. Moreover, $E\{A\bar{B}\}$ is roughly 0, as the asymptotic theory (37) predicts.

CONCLUSIONS. The predictions of stochastic wave theory are in reasonably good agreement with the simulations. Moreover, agreement between the expressions for the total intensity derived using the stochastic and transport theories (i.e. MJ and MJ_s) agree well for small values of $\alpha\tau$ (cf. (40)). However, the overall desired interrelation of these two theories has not been achieved. One anticipates that good agreement should occur for large values of $\alpha\tau$, i.e. in the "deep interior" of thick slabs. For the problem considered though, this regime is precisely where the two theories diverge. Moreover, for large values of $\alpha\tau$, we find that the question of agreement is a moot point since the presence of large intensity fluctuations makes the mean intensity a poor indicator of the actual field behavior within the slab.

In the next section we shall discuss the problem of propagation in a randomly perturbed waveguide. This multimode problem is not fully understood at present and it will be necessary to make very significant simplifying assumptions based on heuristic arguments. On the basis of such qualitative reasoning, however, it seems proper to conclude that our failure to reconcile the stochastic and transport theories for the slab problem stems from the one-dimensional, single-mode character of the problem considered. It appears that one needs many modes, perhaps an infinite set, to put the two theories into consonance.

3. PROPAGATION IN MULTIMODE, RANDOMLY PERTURBED WAVEGUIDES. We now consider a more complex type of wave propagation, encompassing such systems as metal-walled cylindrical waveguides, dielectric waveguides and the ocean, which can be viewed as an acoustic waveguide. The goal is to determine the effect of small random inhomogeneities upon long-range propagation.

A single-frequency $e^{-i2\pi ft}$ time dependence is again assumed. In this case, however, although there exists a preferred direction in which energy is to propagate, the fields are not constant in the remaining two transverse directions. Rather, as the waveguide "traps" the fields and guides them along the preferred direction, they undergo multiple transverse reflections; these transverse reflections, in turn, determine a set of admissable eigenprofiles or modes. The total field can then be viewed as a superposition of these modes, each propagating in the preferred direction with its own characteristic wavenumber.

The metal-walled cylindrical waveguide has a countable infinity of modes. The open waveguides such as the dielectric rod or the ocean have a finite number of bounded or trapped modes and a continuum of radiation modes. These latter radiation modes are used to represent electromagnetic energy radiating away from an optical fiber or acoustic energy penetrating the ocean bottom. The second paper in this sequence will discuss the particular problem of underwater sound propagation in detail. Since our goal at present is to provide background and a heuristic justification for some of the simplifying assumptions, we shall restrict attention to the metal-walled cylindrical guide. The rectangular configuration that we consider is illustrated in Figure 12.

Assume that the guide shown has ideal metallic walls (no

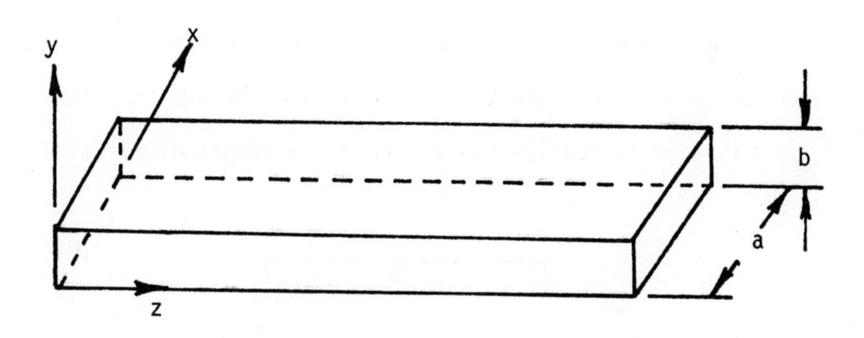

FIGURE 12: RECTANGULAR WAVEGUIDE

geometric imperfections) but that the constitutive parameters of the material filling the guide have small random inhomogeneities. These random imperfections vary in both the transverse and longitudinal directions within the guide; the transverse (i.e. x,y) variation of the inhomogeneities will cause coupling to occur between the modes.

The unperturbed problem has a countable infinity of modes, which can be divided into two classes, E-modes and H-modes. A typical mode is doubly-indexed and has a longitudinal wavenumber

$$\beta_{mn} = \sqrt{k^2 - (m\pi/a)^2 - (n\pi/b)^2}, \qquad (43)$$

$$m,n = 0,1,2,\cdots, \qquad m + n > 0.$$

Thus, the E_{mn} or H_{mn} mode has an $e^{i\beta_{mn}z}$ variation in the direction of propagation. At any given frequency (i.e. fixed value of k), therefore, only a finite number of modes, those with β_{mn} real, will propagate and transport energy. The remaining infinite number of modes will be exponentially decaying in the z-direction; such modes are referred to as evanescent or cut-off modes.

Typically, the random fields associated with the perturbed problem are expanded in the modes of the unperturbed problem. The transverse variation of the random inhomogeneities, however, destroys the orthonormality relations among the modes. They no longer propagate independently of each other; on the contrary, the inhomogeneities couple the modes and a formal substitution of the eigenfunction expansion into the governing equations (i.e. Maxwell's equations) leads to an infinite set of (random) ordinary differential equations for the modal coefficients. This infinite system of equations is then arbitrarily truncated and the resulting finite system of equations is studied.

The details of the modal expansion and the computations leading to the coupled mode equations are extensively discussed in [3] and will be present in the acoustic context in [1]. We shall not require such detail for the present discussion. It suffices to observe that if we consider a segment of randomly perturbed guide of length ℓ illuminated from the left, then the modal expansion and truncation leads to the following problem

$$\frac{d}{dz} X(z) = [D\oplus(-D) + \varepsilon A(z)]X(z), \quad 0 \leqslant z \leqslant \ell, \tag{44}$$

where

$X(z) \equiv 2(M + N) \times 1$ column vector of random modal coupling coefficients,

$$D \equiv \text{diag}\ \{i\beta_1,\cdots,i\beta_M, \quad -\kappa_1,\cdots,-\kappa_N\}, \tag{45}$$

where β_i and κ_i are real positive quantities representing modal propagation and attenuation constants, respectively,

$A(z) \equiv 2(M + N) \times 2(M + N)$ zero mean random modal coupling matrix,

with boundary conditions

$$x_1(0) = \alpha_1, \cdots, x_M(0) = \alpha_M,$$

$$x_{M+1}(0) = x_{M+2}(0) = \cdots = x_{M+N}(0) = 0,$$

$$x_{M+N+1}(\ell) = x_{M+N+2}(\ell) = \cdots = x_{2(M+N)}(\ell) = 0. \quad (46)$$

Here we have assumed a frequency such that M modes can propagate; the system has been truncated so that N evanescent modes are retained in the description. The random matrix A can be assumed to be wide sense stationary and strongly mixing. The boundary conditions reflect the fact that energy is incident from the left. While there are obvious mathematical questions associated with the modal expansion and truncation procedure, we shall simply assume that (44), (46) constitutes the basic problem of interest.

The problem is a stochastic two-point boundary value problem involving a differential equation whose solutions exhibit both exponential growth and decay. We are again interested in the behavior of expectations of physical observables such as intensities, power fluxes, etc. in an appropriate asymptotic limit. There is indirect evidence [29], [30] to the effect that the evanescent modes are negligible. This assumption is usually made ab initia and it seems plausible on physical grounds. However, the formulation of an appropriate limit theorem characterizing (44), (46) in the interior of the inhomogeneous guide segment has not yet been done.

Assume, however, that the evanescent modes are negligible, i.e. that the problem of interest is (44), (46) with $N = 0$. Although progress has been made in studying the input reflection properties of such multimode configurations [31], information about the asymptotic behavior of the fields within the guide has not been developed in detail yet even for the two-mode case.

On the one hand, therefore, the problem appears intractable, particularly in light of the fact that the practical problems of interest involve many modes. On the other hand, though, one might question whether the very presence of a large number of modes might permit additional simplification. This hypothesis is bolstered by the success of theories based upon the forward scattering or parabolic approximation in modelling acoustic and electromagnetic propagation phenomena. The work of F. Tappert and coworkers ([6] and references therein) in the ocean acoustic regime and that of Prokhorov et. al. [5] and Tatarski [4] in the atmospheric electromagnetic regime has produced good agreement with measured data. In both cases, there exists a large number of modes (a continuum of modes in the atmospheric problem). The forward scattering approximation amounts basically to a neglect of the backward travelling waves. Problem (44), (46) with $N = 0$

is further reduced to

$$\frac{d}{dz}\tilde{X}(z) = [\mathrm{diag}\{i\beta_1,\cdots,i\beta_M\} + \varepsilon\tilde{A}(z)]\tilde{X}(z),$$

$$\tilde{X}^T(z) = (x_1(z),\cdots,x_M(z)),$$

$$x_1(0) = \alpha_1,\cdots,x_M(0) = \alpha_M. \tag{47}$$

From a mathematical point of view, we have replaced the two-point boundary value problem (44), (46) by an initial value problem (47). In view of the very pronounced role of backscattering in the one-dimensional slab problem (i.e. the stochastic resonances), this approximation appears somewhat arbitrary and unwarranted even on intuitive grounds. There is, however, some heuristic justification for this forward scattering approximation in the multimode case; we will consider it now.

Recall that in the one-dimensional slab problem, the effective scaled slab length was $\alpha\tau$, where α (defined in (25)) is essentially an evaluation of the power spectrum of the random perturbation μ at wavenumber $2k$. This wavenumber, in turn, represents the difference between the wavenumbers associated with forward propagation $(+k)$ and backward propagation $(-k)$. In this simple problem, the only scatter possible is either forward-to-backward or backward-to-forward. if the random field μ is slowly-varying, so that the power spectrum at $2k$ is small, then the slab will appear relatively transparent to the incident radiation. This fact is consistent with observation that the forward scattering approximation, when applied to (12) with $y = 0$, (i.e. set $B = 0$) leads to the trivial result $|A(x,0,\ell)|^2 = 1$, $0 < x \leqslant \ell$.

In the multimode case, however, there exists a possibility that significant modal interaction can occur and that nontrivial probabilistic effects can emerge in the absence of appreciable backscattering. The situation is illustrated in Figure 13.

For the multimode case, in the context of the forward scattering approximation (i.e. (47)), Papanicolaou [32] has shown that the asymptotic behavior of the modal powers can be described in terms of the following set of coupled power equations.

$$\frac{d}{d\xi} W_m(\xi) = -\sum_{n=1}^{M} a_{mn}[W_m(\xi) - W_n(\xi)], \qquad (48)$$

$$m = 1,\cdots,M, \qquad W_m(0) = |\alpha_m|^2,$$

where

$$\lim_{\substack{\varepsilon\to 0 \\ 0\leqslant\xi\leqslant\xi_0\leqslant\infty}} E\{|x_m(\xi/\varepsilon^2)|^2\} = W_m(\xi). \qquad (49)$$

These equations admit generalization in the acoustic case and will be discussed in [1]. The important feature for the present discussion is the fact that the (nonnegative) coupling coefficients a_{mn} in (48) involve power spectrum evaluations of the matrix elements of $\tilde{A}$ in (47) evaluated at the difference of the mth and the nth longitudinal wavenumbers $|\beta_m - \beta_n|$.

a) PLANE WAVE (ONE-DIM. SLAB)

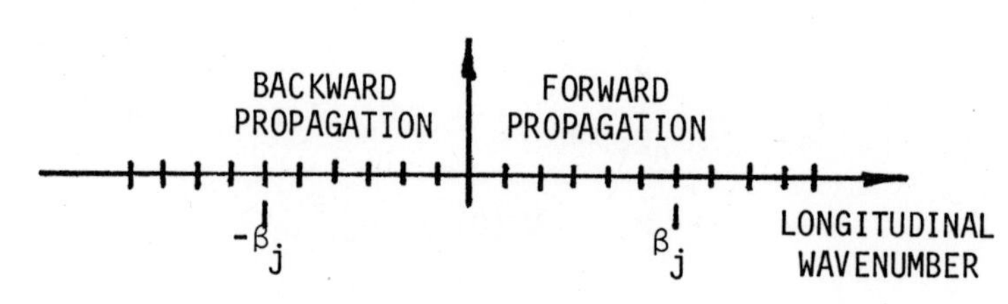

b) MULTIMODE WAVEGUIDE

FIGURE 13: LONGITUDINAL WAVENUMBERS

Summarizing the observations, then, we find that:

i) The one-dimensional random slab exhibits stochastic resonances if the power spectrum of the inhomogeneities is appreciable at the difference wavenumber between forward and backward propagating waves (i.e. $2k$). However, if the parameter α is negligible so that $\alpha\tau$ is very small, the slab will appear transparent to the radiation. In this case (with $\xi = -\tau/2$) the solution can be approximated by the trivial result $MJ = 1$ which results from the forward scattering approximation.

ii) Although the multimode problem has not been fully solved, an analysis based on the forward scattering approximation again points out the dependence of the energy exchange mechanism upon spectral evaluations at modal difference wavenumbers. In the multimode case the forward scattering approximation does not reduce to a triviality; interesting probabilistic effects can emerge in the absence of backscattering.

These two observations lead to the conjecture that the asymptotic behavior of the full multimode problem (46), (47) will likewise involve spectral evaluations at the difference wavenumbers. For coupling between forward and backward modes however, such evaluations will occur at relatively large wavenumbers (i.e. $\beta_n + \beta_m$ rather than $|\beta_n - \beta_m|$). Therefore if the matrix elements of A in (44) which couple forward and backward modal coefficients have spectra with negligible content at the sum wavenumbers, we expect that incident forward propagating energy will be scattered and exchanged principally among the other forward-propagating modes, with only a negligible amount lost to backscattering.

APPENDIX A: LIMIT THEOREM FOR STOCHASTIC EQUATIONS WITH A RAPIDLY VARYING COMPONENT. This appendix presents a theorem and corollary developed by the authors; the theorem is discussed in [33].

Let (Ω,F,P) denote a probability space. We consider the following stochastic initial value problem.

$$\frac{d}{dt}X = \varepsilon F^{(1)}(X,Y,t,\omega) + \varepsilon^2 G^{(1)}(X,Y,t,\omega),$$

$$\frac{d}{dt}Y = - AY + \varepsilon F^{(2)}(X,Y,t,\omega) + \varepsilon^2 G^{(2)}(X,Y,t,\omega),$$

$$X = X(t,s,x,y,\omega,\varepsilon), \qquad Y = Y(t,s,x,y,\omega,\varepsilon),$$

$$X|_{t=s} = x \in R^{n_1}, \qquad Y|_{t=s} = y \in R^{n_2}, \qquad \text{(a-1)}$$

where $A \equiv \text{diag}\{a_1,\cdots,a_{n_2}\}$ with $a_i > 0$, $1 \leqslant i \leqslant n_2$, $\omega \in \Omega$ and ε is a small nonnegative parameter.

Note that in the absence of random perturbation, i.e. $\varepsilon = 0$, X would remain constant while Y would decay exponentially to 0. The problem is studied in the diffusion approximation, an asymptotic limit developed by Stratonovich [8] and Khasminskii [9]. On time scales of order $1/\varepsilon^2$ conditional expectations of functions of the solution process for (a-1) are found to behave asymptotically as the sum of an initial layer term and the solution of a diffusion equation in an n_1-dimensional coordinate space. This collapse in dimensionality to an n_1-dimensional space is not trivial, however, in the sense that the two components in (a-1) do not totally decouple. Rather, the drift coefficient for the limiting diffusion in R^{n_1} is modified by the presence of the rapidly decaying Y process. The slowly-varying X process is found to converge weakly to a Markov diffusion process. Finally, a corollary is formulated which recovers an extended version of Khasminskii's Gauss-Markov limit theorem [34] as a special case. The proofs are similar to those in [16] and will not be presented here.

NOTATION AND DEFINITIONS. Let $E\{\cdot\}$ denote integration with respect to the probability measure P, $|z|$ the Euclidean norm of a vector z, $F \equiv (F^{(1)},F^{(2)})$ and $G \equiv (G^{(1)},G^{(2)})$. We employ the notation

$$D^m f(x,y) \equiv \frac{\partial^m f(x,y)}{\partial x_1^{\beta_1} \cdots \partial x_{n_1}^{\beta_{n_1}} \partial y_1^{\beta_{n_1+1}} \cdots \partial y_{n_2}^{\beta_N}},$$

where β_i is a nonnegative integer, $N \equiv n_1 + n_2$ and $|\beta| \equiv \sum_{i=1}^{N} \beta_i = m$. We use $D^m f$ without reference to the particular underlying N-tuple $(\beta_1,\cdots,\beta_N)$. We also define function spaces

$$C^{k,p}(R^N) \equiv \{f: |f| + \sum_{j=1}^{k} \max_{|\beta|=j} |D^j f| \leqslant C(1 + |x|^p + |y|^p)$$

$$\text{for some integer } p\}.$$

Let $\tau \equiv \varepsilon^2 t$ and set

$$X^{(\varepsilon)}(\tau,\sigma,x,y) \equiv X(\tau/\varepsilon^2,\sigma/\varepsilon^2,x,y,\omega,\varepsilon),$$

$$Y^{(\varepsilon)}(\tau,\sigma,x,y) \equiv Y(\tau/\varepsilon^2,\sigma/\varepsilon^2,x,y,\omega,\varepsilon).$$

We shall use C to denote a constant, independent of ε, x, y, ω, t, whose particular value is unimportant. Similarly, we shall use $\overline{p}$ to denote a nonnegative integer.

Hypotheses.

i) Let F_s^t, $0 \leqslant s \leqslant t \leqslant \infty$, denote a family of σ-algebras in F with $F_{s_1}^{t_1} \subset F_{s_2}^{t_2}$ if $0 \leqslant s_2 \leqslant s_1 \leqslant t_1 \leqslant t_2 \leqslant \infty$. We assume that the family is mixing relative to P as follows.

$$\sup\{\sup |P(A|B) - P(A)|\} = \rho(t) \downarrow 0 \quad \text{as} \quad t \to \infty \quad \text{for all} \quad s \geqslant 0,$$

$$\text{where} \quad A \in F_{s+t}^{\infty}, \qquad B \in F_0^s,$$

and the mixing rate ρ satisfies

$$\int_0^\infty \rho^{1/5}(x)dx < \infty.$$

Conditional probabilities relative to F_0^s are assumed to have a regular version (cf. [15], p. 139).

ii) F and G are mappings of $R^N \times [0,\infty) \times \Omega \to R^N$ which are joint measurable with respect to their arguments and which, for fixed x, y, t, are F_t^t-measurable functions of ω. Moreover, we assume that

$$|F(x,y,t,\omega)| + |G(x,y,t,\omega)| \leq C(1 + |x| + |y|), \qquad \text{(a-2)}$$

$$|D^1F(x,y,t,\omega)| + |D^1G(x,y,t,\omega)| \leq C, \qquad \text{(a-3)}$$

$$|D^iF(x,y,t,\omega)| + |D^iG(x,y,t,\omega)| \leq C(1 + |x|^{\bar{p}} + |y|^{\bar{p}}), \quad i = 2,3, \qquad \text{(a-4)}$$

$$\left|\int_{t_0}^{t_0+T} E\{F(x,y,t,\omega)\}dt\right| \leq \frac{C(1 + |x| + |y|)}{1 + T^4}, \qquad \text{(a-5)}$$

$$\left|\int_{t_0}^{t_0+T} E\{D^1F(x,y,t,\omega)\}dt\right| \leq C, \qquad \text{(a-6)}$$

where t_0 and T are ≥ 0.

iii) Define

$$\hat{F}(x,y,t,\omega) \equiv F(x,y,t,\omega) - E\{F(x,y,t,\omega)\}. \qquad \text{(a-7)}$$

We assume that the following limits exist independently of $t_0 \geq 0$.

$$a^{ij}(x) \equiv \lim_{T\to\infty}\left[\frac{1}{T}\int_{t_0}^{t_0+T}\int_{t_0}^{s} E\{\hat{F}_i^{(1)}(x,0,s,\omega)\hat{F}_j^{(1)}(x,0,\lambda,\omega)\}d\lambda ds\right], \quad i,j = 1,\cdots,n_1, \qquad \text{(a-8)}$$

$$b^i(x) \equiv \lim_{T\to\infty} \left[\frac{1}{T} \int_{t_0}^{t_0+T} \int_{t_0}^{s} \left[\sum_{k=1}^{n_1} E\{\frac{\partial}{\partial x_k} \hat{F}_i^{(1)}(x,0,s,\omega)\hat{F}_k^{(1)}(x,0,\lambda,\omega)\} \right.\right.$$

$$\left. + \sum_{k=1}^{n_2} e^{-a_k(s-\lambda)} E\{\frac{\partial}{\partial y_k} \hat{F}_i^{(1)}(x,0,s,\omega)\hat{F}_k^{(2)}(x,0,\lambda,\omega)\} \right] d\lambda ds$$

$$\left. + \frac{1}{T} \int_{t_0}^{t_0+T} E\{G_i^{(1)}(x,0,s,\omega)\}ds \right] , \quad i = 1,\cdots,n_1. \tag{a-9}$$

Moreover, we shall adopt "rate-of-approach" hypotheses for the limits defined by (a-8), (a-9). We shall assume that the magnitude of the difference between the limit a^{ij} and the integral expression in (a-8) is less than $C(1+|x|^2)/(1+T)$, while the magnitude of the difference between b^i and the sum of integrals in (a-9) is less than $C(1+|x|)/(1+T)$.

iv) In terms of the a^{ij} and b^i we define the following (possibly degenerate) elliptic differential operator on $C^{2,\bar{p}}(R^{n_1})$.

$$Lf(x) \equiv \sum_{i,j=1}^{n_1} a^{ij}(x)\frac{\partial^2}{\partial x_i \partial x_j} f(x) + \sum_{i=1}^{n_1} b^i(x)\frac{\partial}{\partial x_i} f(x). \tag{a-10}$$

We shall assume that for $f \in C^{4,p}(R^{n_1})$, the final value problem

$$\frac{\partial}{\partial\sigma}u(\sigma,\tau,x) + Lu(\sigma,\tau,x) = 0, \qquad \lim_{\sigma\uparrow\tau} u(\sigma,\tau,x) = f(x), \tag{a-11}$$

has a solution $u(\sigma,\tau,x)$ which as a function of x belongs to

$C^{4,\bar{p}}(R^{n_1})$. For this, it is sufficient that the matrix $(a^{ij}(x))$ has a square root $(c^{ij}(x))$ for which the following hold.

$$|b^i(x)| + |c^{ij}(x)| \leq C(1 + |x|), \quad \text{(a-12)}$$

$$|D^1 b^i(x)| + |D^1 c^{ij}(x)| \leq C, \quad \text{(a-13)}$$

$$b^i, c^{ij} \in C^{4,\bar{p}}(R^{n_1}), \quad i,j = 1,\cdots,n_1.$$

Then, Ito's calculus can be used to insure the existence of a solution possessing the desired smoothness and growth properties [13].

We shall continue to denote by u the solution of (a-11) when the final data is replace by $f(x,0)$ (with $f(x,y) \in C^{4,\bar{p}}(R^N)$).

THEOREM. Let hypotheses (i)-(iv) hold. Then there exists an integer $\bar{p}$ such that

$$|E\{f(X^{(\varepsilon)}(\tau,\sigma,x,y),Y^{(\varepsilon)}(\tau,\sigma,x,y))|F_0^{\sigma/\varepsilon^2}\} - [f(x,e^{-A(\tau-\sigma)/\varepsilon^2}y)$$

$$-f(x,0)] - u(\sigma,\tau,x)| \leq \varepsilon^{1/4}C(1 + |x|^{\bar{p}} + |y|^{\bar{p}}),$$

$$0 \leq \sigma \leq \tau \leq \tau_0 < \infty. \quad \text{(a-14)}$$

Let $X^{(0)}(\tau,x)$ denote the n_1-dimensional Markov diffusion process having infinitesimal generator L and satisfying $X^{(0)}(0,x) = x$. Then, as $\varepsilon \to 0$, $X^{(\varepsilon)}(\tau,0,x,y)$ converges weakly to $X^{(0)}(\tau,x)$ for $0 \leqslant \tau \leqslant \tau_0$ and x,y in a compact subset of R^N.

Hypothesis (ii) obviously includes the important special case $E\{F(x,y,t,\omega)\} = 0$. In the context of this special case, this theorem was used to study the input reflection matrix of a waveguide model incorporating both propagating and evanescent modes [30]. The one-dimensional slab problem discussed in Section 2 not only assumes that $E\{F\} = 0$ but that $n_2 = 0$ (i.e. there is no fast-varying component); this problem can be analyzed using the results of [16].

Our aim in developing the theorem in the general context is to simultaneously extend and unify two theorems of Khasminskii [9], [34]. We shall now formulate a corollary which recovers the Gauss-Markov limit theorem [34] as a special case.

COROLLARY. Consider the initial value problem

$$\frac{d}{dt}X = \varepsilon H^{(1)}(X,Y,t,\omega),$$

$$\frac{d}{dt}Y = - AY + \varepsilon H^{(2)}(X,Y,t,\omega), \tag{a-15}$$

$$X(s,s,x,y,\omega,\varepsilon) = x \in R^{n_1},$$

$$Y(s,s,x,y,\omega,\varepsilon) = y \in R^{n_2},$$

$$n_1 + n_2 = N,$$

where the matrix A is the same as in (a-1) and $H \equiv (H^{(1)},H^{(2)})$. Assume that H satisfies (a-2) and (a-3); however, instead of (a-4)-(a-6) we assume that

$$\lim_{T\to\infty} \left| \left[\frac{1}{T} \int_{t_0}^{t_0+T} E\{D^i H(x,y,t,\omega)\}dt \right] \right| = D^i \overline{H}(x,y), \quad i=0,1,2,3, \quad \text{(a-16)}$$

$$\left| \int_{t_0}^{t_0+T} E\{H(x,y,t,\omega) - \overline{H}(x,y)\}dt \right| \leqslant \frac{C(1+|x|+|y|)}{1+T^4}, \quad \text{(a-17)}$$

$$\left| \int_{t_0}^{t_0+T} E\{D^1[H(x,y,t,\omega) - \overline{H}(x,y)]\}dt \right| \leqslant C, \quad \text{(a-18)}$$

where the limits and estimates exist independently of $t_0 \geqslant 0$.

Let $\tau \equiv \varepsilon t$ and let $Z(\tau,\sigma,z^{(1)},z^{(2)},\varepsilon) \equiv (Z^{(1)},Z^{(2)})$ denote the nonrandom trajectory satisfying

$$\frac{d}{d\tau}Z^{(1)} = \overline{H}^{(1)}(Z^{(1)},Z^{(2)}),$$

$$\frac{d}{d\tau}Z^{(2)} = -\frac{1}{\varepsilon}AZ^{(2)} + \overline{H}^{(2)}(Z^{(1)},Z^{(2)}),$$

$$Z^{(i)}\big|_{\tau=\sigma} = z^{(i)}, \quad i = 1,2. \quad \text{(a-19)}$$

Let $X^{(\varepsilon)}(\tau,\sigma,x,y) \equiv X(\tau/\varepsilon,\sigma/\varepsilon,x,y,\omega,\varepsilon)$ and $Y^{(\varepsilon)}(\tau,\sigma,x,y) \equiv Y(\tau/\varepsilon,\sigma/\varepsilon,x,y,\omega,\varepsilon)$. Define

$$W = (W^{(1)},W^{(2)}) \equiv \frac{1}{\sqrt{\varepsilon}}(X^{(\varepsilon)} - Z^{(1)},Y^{(\varepsilon)} - Z^{(2)}). \quad \text{(a-20)}$$

We apply the limit theorem with ε replaced by $\sqrt{\varepsilon}$. The resulting infinitesimal generator is

$$L \equiv \sum_{i,j=1}^{n_1} a^{ij}(z^{(1)}) \frac{\partial^2}{\partial w_i \partial w_j} + \sum_{i,j=1}^{n_1} w_j \frac{\partial}{\partial z_j} \overline{H}_i^{(1)}(z^{(1)},0) \frac{\partial}{\partial w_i}$$

$$+ \sum_{i=1}^{n_1} \overline{H}_i^{(1)}(z^{(1)},0) \frac{\partial}{\partial z_i} , \tag{a-21}$$

where

$$a^{ij}(z^{(1)}) \equiv \lim_{T\to\infty} \left[\frac{1}{T} \int_{t_0}^{t_0+T} \int_{t_0}^{s} E\{\hat{H}_i^{(1)}(z^{(1)},0,s,\omega)\hat{H}_j^{(1)}(z^{(1)},0,\lambda,\omega)\} d\lambda ds \right] ,$$

$$i,j = 1,\cdots,n_1 , \tag{a-22}$$

and $\hat{H} \equiv H - \overline{H}$.

Let $f(w^{(1)},w^{(2)},z^{(1)},z^{(2)}) \in C^{4,p}(R^{2N})$. In applying the theorem, we shall set $w^{(1)} = w^{(2)} = 0$, $z^{(1)} = x$, $z^{(2)} = y$ since we are interested in the case where Z is the "average" trajectory of the $(X^{(\varepsilon)},Y^{(\varepsilon)})$ process. Let $u(\sigma,\tau,w^{(1)},z^{(1)})$ denote the solution of the final value problem

$$\frac{\partial}{\partial\sigma}u + Lu = 0, \quad u(\tau,\tau,w^{(1)},z^{(1)}) = f(w^{(1)},0,z^{(1)},0). \tag{a-23}$$

Then, the counterpart of (a-14), with $W = W(\tau,\sigma,w^{(1)},w^{(2)},\varepsilon)$, is

$$|E\{f(W^{(1)}(\tau,\sigma,0,0,\varepsilon),W^{(2)}(\tau,\sigma,0,0,\varepsilon),Z^{(1)}(\tau,\sigma,x,y,\varepsilon),$$

$$Z^{(2)}(\tau,\sigma,x,y,\varepsilon))|F_0^{\sigma/\varepsilon}\} - [f(0,0,x,e^{-A(\tau-\sigma)/\varepsilon}y) \tag{a-24}$$

$$- f(0,0,x,0)] - u(\sigma,\tau,0,x)| \leqslant \varepsilon^{1/8}C(1+|x|^{\overline{p}}+|y|^{\overline{p}}),$$

$$0 \leqslant \sigma \leqslant \tau \leqslant \tau_0 < \infty.$$

We also find that $(W(\tau,0,0,0,\varepsilon), Z^{(1)}(\tau,0,x,y,\varepsilon))$ converges weakly on $[0,\tau_0]$ as $\varepsilon \downarrow 0$ to the $2n_1$-dimensional Markov diffusion process $(W^{(1)}, Z^{(1)})$ having infinitesimal generator L and initial condition $(0,x)$.

Let $Z^{(1)}$ be the solution of

$$\frac{d}{d\tau} Z^{(1)} = \bar{H}^{(1)}(Z^{(1)},0), \quad Z^{(1)}(0,0,x) = x. \tag{a-25}$$

If $\Phi(\tau,s)$ denotes the fundamental matrix solution of

$$\frac{d}{d\tau}\Phi(\tau,s) = \frac{\partial}{\partial Z^{(1)}}\bar{H}^{(1)}(Z^{(1)}(\tau,0,x),0)\Phi(\tau,s), \qquad \Phi(s,s) = I, \quad 0 \leqslant s \leqslant \tau, \tag{a-26}$$

then $W^{(1)}(\tau,0,0)$ is a Gaussian process with zero mean and covariance matrix

$$E\{W^{(1)}(\tau_1,0,0)W^{(1)T}(\tau_2,0,0)\} = 2\int_0^{\min[\tau_1,\tau_2]} \Phi(\tau_1,s)a(Z^{(1)}(s,0,x))\Phi^T(\tau_2,s)ds. \tag{a-27}$$

This reduces to a result of Khasminskii [34] when $n_2 = 0$ in (a-15).

Appendix B: Derivation of Cross-Correlation Asymptotic Behavior. To illustrate the computations underlying the asymptotic results of Section 2, we sketch the derivation of equation (35).

Assume that $x_2 \geqslant x_1$. The two observation points again divide the slab into three segments $[0,x_1]$, $[x_1,x_2]$, $[x_2,\ell]$ which we refer to as regions 1, 2 and 3, respectively. Noting (19) and (20) we have

$$A(x_2,0,\ell)\bar{A}(x_1,0,\ell) = \frac{\bar{a}_3 a_{32}}{|a_{32}a_1 + b_{32}\bar{b}_1|^2}. \tag{b-1}$$

In terms of the polar coordinate representation (21), the denominator of (b-1) can be expressed as

$$\begin{aligned} |a_{32}a_1 + b_{32}\bar{b}_1|^2 &= \frac{1}{2}[1 + \cosh\theta_1\cosh\theta_{32} \\ &\quad + \cos(\phi_1 + \psi_{32})\sinh\theta_1\sinh\theta_{32}] \\ &\equiv \frac{1}{2}[1 + \cosh\xi]. \end{aligned} \tag{b-2}$$

We now use the relations [35]

$$\frac{2}{1+\cosh\xi} = \int_{-\infty}^{\infty} \frac{\pi t \sin \pi t}{\cosh^2 \pi t} P_\nu(\cosh\xi)dt, \qquad \nu = -\frac{1}{2} + it, \tag{b-3}$$

$$P_\nu(\cosh\xi)$$

$$= \sum_{m=-\infty}^{\infty} \frac{\Gamma(\nu-|m|+1)}{\Gamma(\nu+|m|+1)} e^{im(\phi_1+\psi_{32})} P_\nu^{|m|}(\cosh\theta_1)P_\nu^{|m|}(\cosh\theta_{32}) \tag{b-4}$$

to expand the denominator of (b-1). For brevity, we shall denote the asymptotic limit defined by the left side of (35) by $M\bar{A}A(\eta_1,\eta_2,\tau)$. Then, using (b-2) and (b-3),

$$M\bar{A}A(\eta_1,\eta_2,\tau) = E\Big\{[e^{i(\phi_2+\psi_2)/2}\cosh\frac{\theta_2}{2}\cosh^2\frac{\theta_3}{2}$$

$$+ e^{-i(\phi_2-\psi_2)/2} e^{-i\psi_3} \sinh\frac{\theta_2}{2}\sinh\frac{\theta_3}{2}\cosh\frac{\theta_3}{2}] \int_{-\infty}^{\infty} \frac{\pi t \sinh \pi t}{\cosh^2 \pi t}$$

$$[\sum_{m=-\infty}^{\infty} \frac{\Gamma(\nu-|m|+1)}{\Gamma(\nu+|m|+1)} e^{im(\phi_1+\psi_{32})} P_\nu^{|m|}(\cosh \theta_1) P_\nu^{|m|}(\cosh \theta_{32})]dt\Big\}. \quad \text{(b-5)}$$

In equation (b-5) the indices refer to the independent incre-ents of the limiting matrix-valued Markov diffusion process. Note that the angle ψ_1 does not appear. Therefore, we assume that ϕ_1 is defined on $[0,2\pi)$ and that ψ_1 is defined on $[0,4\pi)$. As [24] indicates, the marginal density for the (ϕ_1,θ_1) process splits into the product of a uniformly distributed ϕ_1 density and the θ_1 density. Performing the ϕ_1 integration singles out the $m = 0$ term in (b-5).

Since only the $m = 0$ term in (b-5) provides a nonzero contribution, and since

$$P_\nu(\cosh \theta_{32}) = P_\nu(\cosh \theta_2 \cosh \theta_3 + \cos(\phi_2+\psi_3)\sinh \theta_2 \sinh \theta_3)$$

$$= \sum_{n=-\infty}^{\infty} \frac{\Gamma(\nu-|n|+1)}{\Gamma(\nu+|n|+1)} e^{in(\phi_2+\psi_3)} P_\nu^{|n|}(\cosh \theta_2) P_\nu^{|n|}(\cosh \theta_3), \quad \text{(b-6)}$$

we note that the angle ϕ_3 does not actually appear in (b-5). Thus, we assume that ψ_3 is defined on $[0,2\pi)$ and ϕ_3 on $[0,4\pi)$. Then angle ψ_3 becomes uniformly distributed. After computing the ψ_3 integration, we have

$$M\overline{AA}(\eta_1,\eta_2,\tau) = E\Big\{\frac{1}{2\sqrt{2}} \int_{-\infty}^{\infty} \frac{\pi t \sinh \pi t}{\cosh^2 \pi t} P_\nu(u_1) e^{i(\phi_2+\psi_2)/2}$$

$$\cdot[(u_2+1)^{1/2}(u_3+1)P_\nu(u_2)P_\nu(u_3) + \frac{\Gamma(\nu)}{\Gamma(\nu+2)}(u_2-1)^{1/2}(u_3^2-1)^{1/2}$$

$$P_\nu^1(u_2)P_\nu^1(u_3)]dt\Big\}, \qquad u_i \equiv \cosh\theta_i, \quad i = 1,2,3. \tag{b-7}$$

In terms of the variable $u = \cosh\theta$

$$VP_\nu(u) = \alpha\frac{\partial}{\partial u}\left[(u^2-1)\frac{\partial}{\partial u}P_\nu(u)\right] = \alpha\nu(\nu+1)P_\nu(u). \tag{b-8}$$

Therefore,

$$E\{P_\nu(u(\tfrac{\tau}{2}+\eta_1))\} = e^{-\alpha(t^2+1/4)(\tau/2+\eta_1)}. \tag{b-9}$$

We also use the relations

$$(u \pm 1)P_\nu(u) = (2\nu+1)^{-1}[(\nu+1)P_{\nu+1}(u) + \nu P_{\nu-1}(u)] \pm P_\nu(u),$$

$$\sqrt{u^2-1}\,P_\nu^1(u) = \frac{\nu(\nu+1)}{2\nu+1}[P_{\nu+1}(u) - P_{\nu-1}(u)] \tag{b-10}$$

to obtain

$$M\bar{A}A(\eta_1,\eta_2,\tau) = E\Big\{\frac{e^{\alpha\tau/4-\alpha(3\eta_2+\eta_1)/4}}{2}\int_{-\infty}^{\infty} e^{-t^2\alpha(\tau-\eta_2+\eta_1)}\frac{\pi t\sinh\pi t}{\cosh^2\pi t}$$

$$\cdot e^{i(\phi_2+\psi_2)/2}[M_+(t,\tfrac{\tau}{2}-\eta_2)\sqrt{\frac{u_2+1}{2}} - N(t,\tfrac{\tau}{2}-\eta_2)\sqrt{\frac{u_2+1}{2}}P_\nu^1(u_2)]dt\Big\}, \tag{b-11}$$

where M_+ and N are defined in (34). Note that the dependence upon the region 2 variables involves u_2 and $\frac{1}{2}(\phi_2 + \psi_2) \equiv \lambda_2$. In terms of these variables, the marginal generator becomes (cf. p. 1744 of [24])

$$V = \alpha\left|\frac{\partial}{\partial u}\left((u^2-1)\frac{\partial}{\partial u}\right) + \frac{1}{4}\left(\frac{u-1}{u+1}\right)\frac{\partial^2}{\partial\lambda^2}\right| + \frac{\gamma}{4}\frac{\partial^2}{\partial\lambda^2} - \frac{\beta}{2}\frac{\partial}{\partial\lambda}. \qquad \text{(b-12)}$$

Expanding the marginal density function for the (u_2,λ_2) process in a Fourier series, we have

$$\tilde{p}(\xi,u,\lambda) \equiv \sum_{n=-\infty}^{\infty} \tilde{p}_n(\xi,u)e^{in\lambda}. \qquad \text{(b-13)}$$

It follows that $\tilde{p}_n$ satisfies

$$\frac{\partial}{\partial\xi}\tilde{p}_n = \alpha\left|\frac{\partial}{\partial u}\left((u^2-1)\frac{\partial}{\partial u}\tilde{p}_n\right) - \frac{n^2}{4}\left(\frac{u-1}{u+1}\right)\tilde{p}_n\right| - n^2\frac{\gamma}{4}\tilde{p}_n + in\frac{\beta}{2}\tilde{p}_n,$$
$$\tilde{p}_n(0,u) = \frac{1}{2\pi}\,\delta(u-1). \qquad \text{(b-14)}$$

To compute expectations of the form appearing in (b-11), (namely of $e^{i\lambda_2}$ times a function of u_2), it suffices to determine $\tilde{p}_{-1}(\xi,u)$. Noting the similarity between (b-14) with $n = 1$ and the equation satisfied by the generalized Legendre function $P^{\nu}_{\frac{1}{2},\frac{1}{2}}$, we use the representation theorem given by Vilenkin ([26], pp. 336-7).

$$\tilde{p}_{-1}(\xi,u) \equiv \frac{1}{4\pi^2}\int_0^{\infty} \Lambda(\xi,\rho)P^{-\frac{1}{2}+i\rho}_{\frac{1}{2},\frac{1}{2}}(u)\rho\coth\pi\rho\, d\rho. \qquad \text{(b-15)}$$

Substitution of this representation into (b-14) leads to

$$\Lambda(\xi,\rho) = \frac{e^{-[\alpha(\rho^2+1/4)+(1/4)(\gamma+\alpha+i2\beta)]\xi}}{2\pi}. \qquad \text{(b-16)}$$

The function $\tilde{p}_{-1}(\eta_2-\eta_1,u)$ is simply referred to as $p(\eta_2-\eta_1,u)$ in (34). Using this function we can complete the computations in (b-11) and obtain (35).

REFERENCES

1. W. Kohler and G. C. Papanicolaou, *Fluctuation in Underwater Sound Propagation, II*, this volume.

2. T. A. Abele, D. A. Alsberg, P. T. Hutchison, *A High Capacity Digital Communication System Using TE_{01} Transmission in Circular Waveguide*, IEEE Trans. MTT, Vol. MTT 23, No. 4, (1975), pp. 326-333.

3. D. Marcuse, *Light Transmission Optics*, Van Nostrand-Reinhold, Princeton, N.J. (1972).

4. V. I. Tatarskii, *The Effects of the Turbulent Atmosphere on Wave Propagation*, NTIS TT-68-50464, Springfield, Virginia.

5. A. M. Prokhorov, F. V. Bunkin, K. S. Gochelashvily and V. I. Shishov, *Laser Irradiance Propagation in Turbulent Media*, Proceedings of IEEE, 63, 5 (1975), pp. 790-811.

6. S. M. Flatté and F. D. Tappert, *Calculation of the Effect of Internal Waves on Oceanic Sound Transmission*, J. Acoustical Soc. of America, 58, 6 (1975), pp. 1151-1159.

7. G. C. Papanicolaou and S. R. S. Varadhan, *A Limit Theorem with Strong Mixing in Banach Space and Two Applications to Stochastic Differential Equations*,

Comm. Pure Appl. Math., 26 (1973), pp. 497-524.

8. R. L. Stratonovich, *Topics in the Theory of Random Noise*, Vols. 1,2, Gordon & Breach, New York (1963).

9. R. Z. Khasminskii, *A Limit Theorem for Solutions of Differential Equations with a Random Right-Hand Side*, Theory Prob. Application, 11 (1966), pp. 390-406.

10. S. Chandrasekhar, *Radiative Transfer*, Dover, New York (1960).

11. Y. N. Barabanenkov. A. G. Vinogradov, Y. A. Kratsov and V. I. Tatarskii, *Applications of Theory of Multiple Scattering to the Radiation Transfer Equation for a Statistically Inhomogeneous Medium*, Radiofisica, 15 (1972), pp. 1852-1860.

12. G. C. Papanicolaou and R. Burridge, *Transport Equations for the Stokes Parameters from Maxwell's Equations in a Random Medium*, J. Math. Phys., 16, 10 (1975), pp. 2074-2085.

13. I. I. Gihman and A. V. Skorohod, *Stochastic Differential Equations*, Springer, Berlin-Heidelberg-New York (1972).

14. J. L. Doob, *Stochastic Processes*, J. Wiley, New York (1953).

15. I. I. Gihman and A. V. Skorohod, *Introduction to the Theory of Random Processes*, W. B. Saunders, Philadelphia, (1969).

16. G. C. Papanicolaou and W. Kohler, *Asymptotic Theory of*

Mixing Stochastic Ordinary Differential Equations, Comm. Pure Appl. Math., 27 (1974), pp. 641-668.

17. A. Schuster, *Radiation through a Foggy Atmosphere*, Astrophysical J., 21, 1 (1905), pp. 1-22.
18. M. E. Gertsenshtein and V. B. Vasil'ev, *Waveguide with Random Inhomogeneities and Brownian motion in the Lobachevsky Plane*, Theory Prob. Appl., 9, 4 (1959), pp. 391-398.
19. M. E. Gertsenshtein and V. B. Vasil'ev, *The Diffusion Equation for Statistically Inhomogeneous Waveguide*, Radio Eng. Electronics, 4 (1959), pp. 74-83.
20. Y. L. Gazaryan, *The One-Dimension Problem of Propagation of Waves in a Medium with Random Homogeneities*, Sov. Phys. JETP, 29 (1969), pp. 996-1003.
21. R. H. Lang, *Probability Density Function and Moments of the Field in a Slab of One-Dimensional Random Medium*, J. Math. Phys., 14, 12 (1973), pp. 1921-1926.
22. D. Marcuse, *Coupled Power Equations for Backward Waves*, IEEE Trans. MTT, Vol. MTT-20 (1972), pp. 541-546.
23. J. Morrison, G. Papanicolaou and J. B. Keller, *Mean Power Transmission through a Slab of Random Medium*, Comm. Pure Appl. Math., 24 (1971), pp. 473-489.
24. W. Kohler and G. Papanicolaou, *Power Statistics for Wave Propagation in One Dimension and Comparison with Radiative Transport Theory*, J. Math. Phys., 14, 12

(1973), pp. 1733-1745.

25. W. Kohler and G. C. Papanicolaou, *Power Statistics for Wave Propagation in One Dimension and Comparison with Radiative Transport Theory, II*, J. Math. Phys., 15, 12 (1974), pp. 2186-2197.

26. N. J. Vilenkin, *Special Functions and the Theory of Group Representations*, 22, Transl. of Math. Monographs, A.M.S., Providence (1968).

27. J. Morrison, *Average Output Power of an Incident Wave Randomly Coupled to a Reflected Wave*, IEEE Trans. MTT, Vol. MTT-22 (1974), pp. 126-130.

28. U. Frisch, C. Froeschle, J.-P. Scheidecker and P. L. Sulem, *Stochastic Resonance in One-Dimensional Random Media*, Phys. Review, A8 (1973), pp. 1416-1421.

29. V. A. Tutubalin, *Multimode Waveguides and Probability Distributions on a Symplectic Group*, Theory Prob. Application, 16, 4 (1971), pp. 631-642.

30. W. Kohler, *Power Reflection at the Input of a Randomly Perturbed Rectangular Waveguide*, SIAM J. Appl. Math., to appear.

31. R. Burridge and G. Papanicolaou, *The Geometry of Coupled Mode Propagation in One-Dimensional Random Media*, Comm. Pure Appl. Math., 25 (1972), pp. 715-757.

32. G. C. Papanicolaou, *A Kinetic Theory for Power Transfer in Stochastic Systems*, J. Math. Phys., 13 (1972)

pp. 1912-1918.

33. G. C. Papanicolaou and W. Kohler, *Asymptotic Analysis of Deterministic and Stochastic Equations with Rapidly Varying Components*, Comm. Math. Phys., 45 (1975), pp. 217-232.

34. R. Z. Khasminskii, *On Stochastic Processes Defined by Differential Equations with a Small Parameter*, Theory Prob. Applications, 11, 2 (1966) pp. 211-228.

35. W. Magnus, F. Oberhettinger and R. P. Soni, *Formulas and Theorems for the Special Functions of Mathematical Physics*, Springer-Verlag, New York (1966).

Virginia Polytechnic Institute and State University
Courant Institute of Mathematical Sciences, New York University.

FLUCTUATION PHENOMENA IN UNDERWATER SOUND PROPAGATION, II.

BY

WERNER KOHLER

AND

GEORGE C. PAPANICOLAOU*

ABSTRACT. We continue Part I of this paper with a discussion, specifically, of the underwater sound problems and the derivation of couple power equations. Then we show how to formally obtain a transport equations for the Wigner distribution. In two appendices we state and prove two theorems for stochastic differential equations that compliment results of I.

1. INTRODUCTION.

This is a continuation of "Fluctuation Phenomena in Underwater Sound Propagation, I" by the authors. All references herein are included at the end of Part I and are identified by the same numbers.

In Section 2 we begin with the formulation of the underwater sound problem. We emphasize those aspects of the problem that have bearing on the analysis of fluctuations. We consider only time harmonic excitation. General time dependence requires some additional considerations. In Section 3 we show how one

*This work was partially completed at the Applied Mathematics Summer Institute, 1975, which is supported by the ONR under Contract No. N00014-75-C-0921.

can derive coupled power equations. The calculations are formal but can be made rigorous by extending slightly existing results. Power fluctuations can also be computed but we do not do so here.

Section 4 deals with a problem of more general interest: derivation of transport theory from stochastic wave equations. The formal calculations we present are intended for motivation purposes only. We do not know at present how to establish mathematical results of this form.

In the Appendices we state and prove two theorems that compliment results in Appendix A of Part I and in [13], [16], [33] and [34]. They are not related directly to underwater sound propagation.

2. The Underwater Sound Problem.

Let $p(r,\theta,z)$ denote the sound pressure field in cylindrical coordinates with z measured downward from the surface of the ocean $(z = 0)$ and with the time factor $e^{-i\omega t}$ omitted throughout. Let $r \geq 0$ denote the range measured from a point source located at $r = 0$ at depth z_0 and let θ denote the azimuthal angle.

The pressure field satisfied the equation

$$\frac{\partial^2 p}{\partial r^2} + \frac{1}{r}\frac{\partial p}{\partial r} + \frac{1}{r^2}\frac{\partial^2 p}{\partial \theta^2} + \frac{\partial^2 p}{\partial z^2} + k^2[n^2(z) + \varepsilon\mu(r,z)]p = \frac{\partial(r)}{2\pi r}\,\delta(z - z_0),$$

$$r \geq 0, \quad 0 \leq \theta < 2\pi, \quad 0 \leq z < \infty, \quad p(r,\theta,0) = 0. \qquad (2.1)$$

Here, $k = \omega/c_0$ is the nominal wavenumber with c_0 ($\simeq$1.4 km/sec) a nominal sound speed and $n(z)$ denotes the mean index of refraction, $n(z) = c_0/c(z)$, where $c(z)$ is the mean velocity profile. It is assumed that $c(z)$ is a function of depth only. The fluctuations about the means are denoted by $\varepsilon\mu(r,z)$; they are random and they can vary with depth and range. For simplicity, we have assumed that they do not depend on the azimuthal angle θ. The general case can be treated without additional complications.

The fluctuations are multiplied by a small parameter ε and typically $\varepsilon \simeq 10^{-3}$. In addition,

$$\langle\mu(r,t)\rangle = 0, \tag{2.2}$$

where $\langle\cdot\rangle$ denotes expectation (ensemble average). Inside the ocean floor, i.e. for z large enough (larger than 5-6 km.), we assume that the fluctuations μ vanish (or become very weak). The ocean floor is modeled by a discontinuity in the mean velocity profile $c(z)$. Typically, $c(z)$ has a minimum at about one km. from the surface and then increases linearly with depth. At 5 or 6 km., $c(z)$ increases discontinuously to a new value comparable to the nominal speed in the solid earth. This is merely a simple model of what is a very complicated affair: the nature of the ocean floor and its sound propagation properties.

The fluctuations $\mu(r,z)$ are assumed to be statistically stationary in range but not in depth. Correlation lengths in range are generally much larger than correlation lengths in depth. On the other hand we are interested in calculating statistical quantities far away from the source (hundreds of km.

from the source). It is clear that (i) even at low frequencies (20-100 Hz) we are many wavelengths away from the source, (ii) we are many correlation lengths away from the source (one horizontal correlation length is about 500 meters, say) and (iii) correlation length and wave length are of order one relative to each other. We shall therefore compute approximately statistical quantities of interest when ε is small and r is large $r \simeq 1/\varepsilon^2$. In this regime stochastic effects are fully developed and can be described in a relatively simple manner.

We shall now prepare successively problem (2.1) for a stochastic analysis that follows in Section 3.

In view of the enforced azimuthal symmetry, the pressure field is only a function of r and z and it satisfies the simpler equation

$$\frac{\partial^2 p}{\partial r^2} + \frac{1}{r}\frac{\partial p}{\partial r} + \frac{\partial^2 p}{\partial z^2} + k^2[n^2(z) + \varepsilon\mu(r,z)]p = \frac{\delta(r)}{2\pi r}\,\delta(z - z_o)$$

$$0 \leqslant r < \infty, \quad 0 \leqslant z < \infty, \quad p(r,0) = 0. \qquad (2.3)$$

We introduce next the assumption that stochastic effects manifest themselves entirely within the cylindrically spreading regime. This means that there is a region around the source large enough so that the emitted spherical waves have reached their asymptotic cylindrically spreading state. At the same time, this region is small enough so that stochastic effects have not accumulated and can be ignored. One can analyze this (essentially deterministic) matching-to-the-source problem completely, but we shall not do it here. We assume that we may replace p by $p/\sqrt{r}$ (so that the new p has the spreading factor removed) and (neglecting a near field term) the new p satisfies

$$\frac{\partial^2 p}{\partial r^2} + \frac{\partial^2 p}{\partial z^2} + k^2[n^2(z) + \varepsilon\mu(r,z)]p = 0,$$
$$r > 0, \quad z \geqslant 0, \quad p(r,0) = 0. \tag{2.4}$$

In (2.4) we have set the right side of the equation equal to zero. Of course, as it stands (2.4) has solution $p \equiv 0$. However, at $r = 0$ we shall prescribe p in a manner compatible with the matching-to-the-source procedure as well as the forward scattering approximation which we shall now describe.

Consider the differential operator

$$L = \frac{d^2}{dz^2} + k^2 n^2(z), \qquad z > 0,$$
$$\text{Boundary condition zero at } z = 0. \tag{2.5}$$

This operator is self-adjoint in $L^2[0,\infty)$ for a broad class of indices of refraction corresponding to $c(z)$ for the ocean. In general, L has a finite number of eigenvalues and a continuous spectrum. We assume that

$$L\nu_p(z) = \beta_p^2 \nu_p(z), \quad \nu_p(0) = 0, \quad p = 1,2,\cdots,N, \quad (N = N(k)), \tag{2.6}$$

$$L\nu(z,\gamma) = \gamma\nu(z,\gamma), \quad \nu(0,\gamma) = 0, \quad -\infty \leqslant \gamma \leqslant k^2, \tag{2.7}$$

$$(\nu_p,\nu_q) = \int_0^\infty \nu_p(z)\nu_q(z)dz = \delta_{pq}, \qquad (2.8)$$

$$(\nu_p,\nu(\gamma)) = 0, \quad (\nu(\gamma),\nu(\gamma')) = \delta(\gamma - \gamma'), (2.9)$$

where $\nu_p = \nu_p(z)$ and $\nu(\gamma) = \nu(z,\gamma)$ with the z dependence omitted. Note that $N = N(k)$ and, as $k \to \infty$, $N(k)$ increases, i.e. as the frequency ω increases the number of trapped (or propagating) modes $N(k)$ increases. The ocean is a dispersive medium since different modes (wave profiles in the z-variable) travel at different speeds (group velocities) $c_0[\partial\beta_p(k)/\partial k]^{-1}$.

The continuum or radiation modes in the range $0 \leqslant \gamma \leqslant k^2$ correspond to radiation into the bottom of the ocean (i.e. angles of incidence between tangential and critical angle). Radiation modes with $-\infty < \gamma < 0$ correspond to evanescent modes.

Let us expand the solution $p(r,z)$ of (2.4) in terms of the eigenfunctions of L.

$$p(r,z) = \sum_{p=1}^{N} c_p(r)\gamma_p(z) + \int_{-\infty}^{k^2} c(\gamma,r)\nu(z,\gamma)d\gamma. \qquad (2.10)$$

The coefficient functions $c_p(r)$ and $c(\gamma,r)$ are called the *mode amplitudes*. They are complex-valued random functions of the range r and they satisfy the equations

$$\frac{d^2c_p(r)}{dr^2} + \beta_p^2 c_p(r) + \varepsilon k^2 \sum_{q=1}^{N} \hat{\mu}_{pq}(r)c_q(z) + \varepsilon k^2 \int_{-\infty}^{k^2} \hat{\mu}_{p\gamma}(r)c(\gamma,r)d\gamma = 0,$$

$$p = 1,2,\cdots,N, \qquad (2.11)$$

$$\frac{d^2c(\gamma,r)}{dr^2} + \gamma c(\gamma,r) + \varepsilon k^2 \sum_{q=1}^{N} \hat{\mu}_{\gamma q}(r)c_q(r)$$
$$+ \varepsilon k^2 \int_{-\infty}^{k^2} \hat{\mu}_{\gamma\gamma'}(r)c(\gamma',r)d\gamma' = 0,$$
$$-\infty < \gamma \leqslant k^2. \qquad (2.12)$$

These equations are obtained by inserting (2.10) into (2.4) and using (2.8), (2.9). The coefficients in (2.12) are given by†

$$\hat{\mu}_{pq}(r) = (\mu(r)\nu_q,\nu_p) = \int_0^\infty \mu(r,z)\nu_p(z)\nu_q(z)dz,$$

$$\hat{\mu}_{\gamma q}(r) = (\mu(r)\nu_q,\nu(\gamma)),$$

$$\hat{\mu}_{p\gamma}(r) = (\mu(r)\nu(\gamma),\nu_p),$$

$$\hat{\mu}_{\gamma\gamma'}(r) = (\mu(r)\nu(\gamma'),\nu(\gamma)), \qquad (2.13)$$

which are just the "matrix" elements of μ relative to the modes.

We shall assume in the following that evanescent continuous modes $\nu(\gamma,z)$, $-\infty < \gamma < 0$, can be neglected. This is reasonable because these waves do not propagate energy over large distances.

Let us write the mode amplitudes in the form

$$c_p(r) = \frac{1}{\sqrt{\beta_p}}[c_p^+(r)e^{i\beta_p r} + c_p^-(r)e^{-\beta_p r}], \quad p = 1,2,\cdots,N,$$

$$c(\gamma,r) = \gamma^{-1/4}[c^+(\gamma,r)e^{i\sqrt{\gamma}r} + c^-(\gamma,r)e^{-i\sqrt{\gamma}r}], \quad 0 \leqslant \gamma \leqslant k^2. \quad (2.14)$$

The complex-valued random functions $c_p^{\pm}(r)$ and $c^{\pm}(\gamma,r)$ are

†Note however that μ vanishes for large z by hypothesis.

called the *forward* (with +) and *backward* (with -) *mode amplitudes*. This is consistent with the assumed time factor $e^{-i\omega t}$. The factors $\beta_p^{-1/2}$ and $\gamma^{-1/4}$ are introduced in (2.14) in order that some formulas below become symmetric. In addition to the definition (2.14), we prescribe also the following relationship between $c_p^{\pm}$ and $c^{\pm}(\gamma)$.

$$e^{i\beta_p r}\frac{dc_p^+(r)}{dr} + e^{-i\beta_p r}\frac{dc_p^-(r)}{dr} = 0,$$

$$e^{i\sqrt{\gamma}r}\frac{dc^+(\gamma,r)}{dr} + e^{-i\sqrt{\gamma}r}\frac{dc^-(\gamma,r)}{dr} = 0. \qquad (2.15)$$

We next insert (2.14) into (2.11) and (2.12) and use (2.15). This way we obtain coupled equations for $c_p^{\pm}(r)$ and $c^{\pm}(\gamma,r)$ which involve only first order derivatives in r. We *assume* that $c_p^-(r)$ and $c^-(\gamma,r)$ can be set to zero (can be neglected) in these equations. This is the *forward scattering approximation*. Its justification in the context of the stochastic problem rests mostly on evidence (numerical and experimental) that it is acceptable. It is believed to be a very good approximation for the underwater sound problem for a broad range of frequencies.

We now write the coupled mode equations for the *forward mode amplitudes* c_p^+ and $c^+(\gamma)$ and from now on we shall omit the superscript +.

$$\frac{dc_p(r)}{dr} = i\varepsilon \sum_{q=1}^{N} \mu_{pq}(r) e^{i(\beta_q-\beta_p)r} c_q(r)$$

$$+ i\varepsilon \int_0^{k^2} \mu_{p\gamma}(r) e^{i(\sqrt{\gamma}-\beta_p)r} c(\gamma,r)d\gamma, \quad p = 1,2,\cdots,N,$$

$$\frac{dc(\gamma,r)}{dr} = i\varepsilon \sum_{q=1}^{N} \mu_{\gamma q}(r) e^{i(\beta_q-\sqrt{\gamma})r} c_q(r)$$

$$+ i\varepsilon \int_0^{k^2} \mu_{\gamma\gamma'}(r) e^{i(\sqrt{\gamma'}-\sqrt{\gamma})r} c(\gamma',r)d\gamma', \quad 0 < \gamma \leqslant k^2. \tag{2.16}$$

Here we have introduced the notation

$$\mu_{pq}(r) = \frac{k^2}{2\sqrt{\beta_p\beta_q}} \hat{\mu}_{pq}(r), \quad \mu_{p\gamma}(r) = \frac{k^2}{2\sqrt{\gamma}} \hat{\mu}_{p\gamma}(r),$$

$$\mu_{\gamma' p}(r) = \frac{k^2}{2\sqrt{\beta_p\sqrt{\gamma'}}} \hat{\mu}_{\gamma' p}(r),$$

$$\mu_{\gamma'\gamma}(r) = \frac{k^2}{2(\gamma\gamma')^{1/4}} \hat{\mu}_{\gamma'\gamma}(r). \tag{2.17}$$

Note that the coefficients μ as arranged in (2.17) form a real symmetric "matrix" allowing for continuously indexed quantities in the obvious way.

We must now assign *initial values* at $r = 0$ to the system (2.16). This brings us back to the problem of matching-to-the-source (Eq. (2.4)) and the forward scattering approximation. We assume that by side calculations (which are deterministic) we have found "effective" initial values

$$c_p(0) = c_{po}, \quad p = 1,2,\cdots,N,$$
$$c(\gamma,0) = c_o(\gamma), \quad 0 \leq \gamma \leq k^2, \tag{2.18}$$

where c_{po} and $c_o(\gamma)$ are complex numbers. These numbers characterize the nature of the source, i.e. the manner in which the source transfers energy into the trapped modes and into the radiation modes.

It is convenient to write (2.16) in matrix form. We must allow however, in the usual way, for continuously varying indices γ,γ' in $[0,k^2]$. With this convention we write

$$\mu(r) = \begin{pmatrix} \mu_{pq}(r) & \mu_{p\gamma}(r) \\ & \\ \mu_{\gamma'q}(r) & \mu_{\gamma\gamma'}(r) \end{pmatrix}, \tag{2.19}$$

with entries defined by (2.17). The matrix $\mu(r)$ is symmetric and real valued and it is also a random process. If we also introduce the mode-amplitude vector

$$c(r) = \begin{bmatrix} c_1(r) \\ \vdots \\ c_N(r) \\ \vdots \\ c(\gamma,r) \\ \vdots \end{bmatrix}, \tag{2.20}$$

and the diagonal matrix

$$\beta = \operatorname{diag}\left\{\beta_1, \beta_2, \cdots, \beta_N, \cdots, \sqrt{\gamma}, \cdots\right\}, \qquad 0 \leqslant \gamma \leqslant k^2, \qquad (2.21)$$

then we may write (2.16) compactly as follows.

$$\frac{dc(r)}{dr} = i\varepsilon e^{-i\beta r}\mu(r)e^{i\beta r}c(r), \quad r > 0,$$
$$c(0) = c_0. \qquad (2.22)$$

We shall refer to (2.22) as the *coupled mode equations*. They are an infinite system of coupled stochastic equations in slowly varying form, i.e. with ε multiplying the right hand side and with

$$\langle\mu(r)\rangle = 0, \qquad (2.23)$$

which follows from (2.2). Note also that the total forward propagating energy

$$\sum_{p=1}^{N} |c_p(\varepsilon)|^2 + \int_0^{k^2} |c(\gamma,r)|^2 d\gamma \qquad (2.24)$$

is preserved by (2.22), i.e. it is a quantity independent of the range r and equal to the initial total energy.

3. COUPLED POWER EQUATIONS. In this section we shall analyze the stochastic system (2.22) that governs the evolution of the mode amplitudes of the forward propagating and radiation modes as functions of the range. In the absence of radiation modes, the calculations that follow are fully rigorous under reasonable conditions (cf. [16]). With radiation present, additional considerations are necessary but not substantially new

ones. The reason for this is that the final results are relatively easy to describe (and in a finite dimensional space) even though (2.22) is a complicated system because of the radiation modes.

The first step in the analysis is to introduce the scaled range

$$r = \tau/\varepsilon^2, \tag{3.1}$$

and put

$$\begin{aligned} c_p^{\varepsilon}(\tau) &= c_p(\tau/\varepsilon^2), \quad p = 1,2,\cdots,N, \\ c^{\varepsilon}(\gamma,\tau) &= c(\gamma,\tau/\varepsilon^2), \quad 0 \leq \gamma \leq k^2. \end{aligned} \tag{3.2}$$

In vector notation (with the convention of continuous indices) the scaled form of (2.22) is

$$\begin{aligned} \frac{dc^{\varepsilon}(\tau)}{d\tau} &= \frac{i}{\varepsilon} e^{-i\beta\tau/\varepsilon^2} \mu(\tau/\varepsilon^2) e^{i\beta\tau/\varepsilon^2} c^{\varepsilon}(\tau), \quad \tau > 0, \\ c^{\varepsilon}(0) &= c_0. \end{aligned} \tag{3.3}$$

It is instructive now to carry out the analysis without radiation so that (3.3) has the following form.

$$\begin{aligned} \frac{dc_p^{\varepsilon}(\tau)}{d\tau} &= \frac{i}{\varepsilon} \sum_{q=1}^{N} e^{i(\beta_q-\beta_p)\tau/\varepsilon^2} \mu_{pq}(\tau/\varepsilon^2) c_q(\tau), \quad \tau > 0, \\ c_p(0) &= c_{op}, \qquad p = 1,2,\cdots,N. \end{aligned} \tag{3.4}$$

The matrix process $(\mu_{pq}(\tau))$ is real, symmetric, stationary with mean zero and we define correlations as follows.[†]

$$R_{pq,p'q'}(\sigma) = E\{\mu_{pq}(\sigma+\tau)\mu_{p'q'}(\tau)\}. \qquad (3.5)$$

Let us also assume that the processes $(\mu_{pq}(\tau))$ are mixing (cf. for example [16]) and that

$$\{\beta_p\} \text{ are rationally independent.} \qquad (3.6)$$

Note also that because of the conservation of (2.24) (with no radiation term now) we have that

$$(c_p^{\varepsilon}(\tau)) \text{ is a process on the unit sphere in } C^N, \qquad (3.7)$$

where we assume that $\sum_{p=1}^{N} |c_{op}|^2 = 1$.

The existing theory ([9], [16]) leads now to the following result.

The process $(c_p^{\varepsilon}(\tau))$ converges weakly[††] as $\varepsilon \downarrow 0$, $0 \leqslant \tau \leqslant T < \infty$ to a diffusion Markov process $(c_p(\tau))$ whose infinitesimal generator L is given by (3.13) below.

We note the formulas that follow are obtained by direct application of formulas in [16], for example, specifically (2.27)-(2.37) (pp. 646-649).

To describe the generator L of the limit process

[†]We use $<\cdot>$ or $E\{\cdot\}$ for expectation.

[††]As a probability measure on $C([0,T];C^N)$ (which turns out to be concentrated on the unit sphere in C^N).

$(c_p(\tau))$ we proceed as follows. Since $c_p(\tau)$, $p = 1,\cdots,N$, is complex valued, it is convenient to work with c_p and c_p^* (the complex conjugate) rather than real and imaginary parts. If $f(c,c^*)$ is a real function of a complex variable $c = x + iy$, $c^* = x - iy$ then, as usual,

$$\frac{\partial f}{\partial c} = \frac{1}{2}\left(\frac{\partial f}{\partial x} - i\,\frac{\partial f}{\partial y}\right), \qquad \frac{\partial f}{\partial c^*} = \frac{1}{2}\left(\frac{\partial f}{\partial x} + i\,\frac{\partial f}{\partial y}\right). \tag{3.8}$$

Let

$$\tilde{a}_{pq} = \int_{-\infty}^{\infty} R_{pp,qq}(s)ds, \tag{3.9}$$

$$a_{pq} = \int_{-\infty}^{\infty} R_{pq,pq}(t)\cos(\beta_p-\beta_q)t dt, \tag{3.10}$$

$$\hat{a}_{pq} = \int_{0}^{\infty} R_{pq,pq}(t)\sin(\beta_p-\beta_q)t dt, \tag{3.11}$$

$$A_{pq} \equiv c_p \frac{\partial}{\partial c_q} - c_q^* \frac{\partial}{\partial c_p^*}, \qquad A_{pq}^* = -A_{qp}. \tag{3.12}$$

Then, the infinitesimal generator L of $(c_p(\tau))$ has the form

$$Lf = \sum_{1\leq q<p\leq N}\left\{\frac{1}{2}\, a_{pq}(A_{pq}A_{pq}^*+A_{pq}^*A_{pq}) + \tilde{a}_{pq}A_{pp}A_{qq}^* + i\hat{a}_{pq}(A_{qq}-A_{pp})\right\}f$$
$$+ \frac{1}{2}\sum_{p=1}^{N}\tilde{a}_{pp}A_{pp}A_{pp}^* f. \tag{3.13}$$

A number of interesting conclusions can be drawn from formula (3.13) for the limiting generator.

First, if we let

$$\lim_{\varepsilon\downarrow 0} E\{|c_p^{\varepsilon}(\tau)|^2\} = W_p(\tau), \tag{3.14}$$

then we find from (3.13) that $(a_{pq} = a_{qp})$

$$\frac{dW_p(\tau)}{d\tau} = \sum_{q=1}^{N} (a_{pq}W_q - a_{qp}W_p), \quad \tau > 0,$$

$$W_p(0) = c_{op}c^*_{op}, \quad p = 1,2,\cdots,N. \tag{3.15}$$

Thus, the expectation of the modulus squared of the propagating mode amplitudes, the *propagating mode power amplitudes*, converges in the limit $\varepsilon \to 0$ to the solution of the *transport equation* (3.15). This is quite an interesting conclusion because (3.15) makes a lot of physical sense, and it is repeatedly derived by phenomenological considerations. Note that the approach we have followed, i.e. the limit theorem, yields formulas for the transport coefficients a_{pq} (cf. (3.10)). They are the power spectra of the coupling coefficients μ_{pq} in (3.4) at the difference frequencies.

Of course, (3.13) contains much more information than merely (3.15). For example, one gets closed equations also for quantities of the form

$$\lim_{\varepsilon \downarrow 0} E\left\{|c^{\varepsilon}_{p_1}(\tau)|^{\alpha_1}|c^{\varepsilon}_{p_2}(\tau)|^{\alpha_2} \cdots |c^{\varepsilon}_{p_m}(\tau)|^{\alpha_m}\right\}, \tag{3.16}$$

in particular for *mode power amplitude fluctuations*, etc. Note also that only the constants a_{pq} of (3.10) enter in the description of the limit for expressions of the form (3.16).

Let us consider now the case *with radiation*, i.e. (3.3)

in full. The usual theorems do not apply except in a formal way as was pointed out at the beginning of this section. However, we expect that the following result holds under quite general conditions, in particular the ones of Section 2 that led to (3.3).

The propagating mode amplitudes $(c_p^\varepsilon(\tau))$ converge weakly as $\varepsilon \to 0$, $0 \leq \tau \leq T < \infty$, to the diffusion Markov process $(c_p(\tau))$ on C^N (not the unit sphere in C^N) whose generator L^R is given by

$$L^R f = Lf - \sum_{p=1}^{N} \int_0^{k^2} d\gamma \left(\hat{b}_{p\gamma} c_p \frac{\partial}{\partial c_p} + \hat{b}_{p\gamma}^* c_p^* \frac{\partial}{\partial c_p^*} \right), \quad (3.17)$$

where L is given by (3.13) and

$$\hat{b}_{p\gamma} = \int_0^\infty R_{p\gamma,p\gamma}(t) e^{i(\beta_p - \sqrt{\gamma})t} dt, \qquad p = 1,2,\cdots,N,$$

$$0 \leq \gamma \leq k^2. \quad (3.18)$$

In particular, this result implies that $W_p(\tau)$, defined as in (3.14), satisfies

$$\frac{dW_p(\tau)}{d\tau} = -b_p W_p(\tau) + \sum_{q=1}^{N} (a_{pq} W_q - a_{qp} W_p), \quad \tau > 0,$$

$$W_p(0) = |c_{op}|^2, \qquad p = 1,2,\cdots,N, \quad (3.19)$$

$$b_p = 2 \operatorname{Re}\left\{ \int_0^{k^2} \hat{b}_{p\gamma} d\gamma \right\}$$

$$= \int_0^{k^2} d\gamma \int_{-\infty}^{\infty} dt \, R_{p\gamma,p\gamma}(t) \cos(\beta_p - \sqrt{\gamma})t. \quad (3.20)$$

Note that b_p as defined by (3.20) are non-negative constants and are rates with which the propagating modes lose energy to the radiation modes.

Again, as we mentioned above, the fact that (3.19) holds in the limit $\varepsilon \to 0$ for the propagating mode amplitudes makes good physical sense for the underwater sound problem. In the small-fluctuation long-range limit modes exchange power according to (3.19) and they lose some which escapes into the bottom of the ocean.

Of course, much more information can be extracted from (3.17): power fluctuation equations, pulse problems, etc. In addition, one can study the behavior of solutions to (3.19) when the a_{pq} are small, except where $|p-q| \leqslant 1$ and when N is large (diffusion approximation). Such approximations are important when k, the wave number (or the frequency ω), is large since $N = N(k) \uparrow \infty$ as $k \uparrow \infty$. We examine these questions in a forthcoming publication.

4. THE WIGNER DISTRIBUTION. The purpose of this section is to present briefly a derivation of a result given in [11] that shows the formal connections between the asymptotics for stochastic wave equations and radiative transport theory The level of presentation here is comparable to the one in [11]. We do not know, at present, how to give a full mathematical

treatment for the problem that follows.[†]

Let $V(x)$, $x \in R^n$, be a real valued stationary random field with mean zero and consider solutions $u(x)$ of the wave equation

$$\Delta u(x) + \kappa^2(1 + V(x)u(x)) = 0. \qquad (4.1)$$

Naturally, we want solutions that correspond to waves in a region $\mathcal{D}$ with given incident waves on $\partial\mathcal{D}$. However, we shall assume that (4.1) holds in all of space since boundaries are expected to cause additional difficulties.

With any solution $u(x)$ we associate the function

$$w(k,x) = \int e^{ik\cdot x} u(x + \tfrac{y}{2})u^*(x - \tfrac{y}{2})dy, \qquad (4.2)$$

where $*$ is complex conjugate, k is a vector in R^n and the integration is over all space. The function $w(k,x)$ is called the Wigner distribution and it is seen immediately that

$$w^*(k,x) = w(k,x), \qquad (4.3)$$

so w is real. It is not in general non-negative. If $\hat{u}(p)$, $p \in R^n$, denotes the Fourier transform of $u(x)$,

$$\hat{u}(p) = \int e^{ip\cdot x} u(x)dx, \qquad (4.4)$$

[†]Many preliminary steps below can be made rigorous in a more or less standard way. However, the key steps remain formal so we abandon attempts at rigor from the start.

then we obtain from (4.2) the representation

$$w(k,x) = \frac{2^{1-n}}{(2\pi)^n} \int e^{-ip\cdot x}\hat{u}(k + \tfrac{p}{2})\hat{u}^*(k - \tfrac{p}{2})dp. \qquad (4.5)$$

The reason why we introduce the Wigner distribution is that it is a quadratic field quantity that relates to quantities of interest directly. For example, we easily verify that

$$\frac{1}{(2\pi)^n} \int w(k,x)dk = |u(x)|^2, \qquad (4.6)$$

$$\frac{1}{(2\pi)^n} \int kw(k,x)dk = \frac{1}{2i}(u(x)\nabla_x u^*(x) - u^*(x)\nabla_x u(x)), \qquad (4.7)$$

$$\int w(k,x)dx = 2^{1-n}|\hat{u}(k)|^2, \qquad (4.8)$$

$$\int xw(k,x)dx = \frac{2^{1-n}}{2i}\left[\hat{u}^*(k)\nabla_k\hat{u}(k) - \hat{u}(k)\nabla_k\hat{u}^*(k)\right]. \qquad (4.9)$$

This shows that the local energy and energy flux in physical and wavenumber space relate simply to moments of the Wigner distribution.

By direct formal computation we find that if $u(x)$ satisfies (4.1), then $w(k,x)$ satisfies the equation

$$k\cdot\nabla_x w(k,x) = -\frac{-i\kappa^2 2^{1-n}}{(2\pi)^n} \int \hat{V}(2k-2k')e^{-2i(k-k')\cdot x}$$

$$\cdot\ w(k',x)dk' + \frac{i\kappa^2 2^{1-n}}{(2\pi)^n} \int \hat{V}^*(2k-2k')$$

$$\cdot\ e^{2i(k-k')\cdot x}w(k',x)dk'. \qquad (4.10)$$

Here, $\hat{V}(k)$ is the Fourier transform of the random field $V(x)$ so that

$$V(x) = \int e^{ik\cdot x}\hat{V}(k)dk. \tag{4.11}$$

Since $V(x)$ is stationary, its correlation function is

$$E\{V(x + y)V(x)\} = R(y), \tag{4.12}$$

and we have

$$E\{\hat{V}(k)\hat{V}(k')\} = \delta(k + k')\hat{R}(k). \tag{4.13}$$

Note that (4.10) has the appearance of a transport equation with random, complex-valued, scattering kernel.

Next we scale the problem. Let $\varepsilon > 0$ be a small parameter characterizing the size of the random inhomogenities. We replace $V(x)$ by $\varepsilon^{-1}V(x/\varepsilon^2)$ in (4.1) which corresponds to the scaling we usually use (cf. previous sections and Part I). Then the corresponding Wigner distribution $w^{\varepsilon}(k,x)$ satisfies the equation $(c_n = \kappa^2 2^{1-n}/(2\pi)^n)$

$$\begin{aligned} k\cdot\nabla_x w(k,x) = &-\frac{ic_n}{\varepsilon}\int \hat{V}(2k-2k')e^{-2i(k-k')\cdot x/\varepsilon^2} \\ &\cdot w^{\varepsilon}(k',x)dk' + \frac{ic_n}{\varepsilon}\int \hat{V}^*(2k-2k') \\ &\cdot e^{2i(k-k')\cdot x/\varepsilon^2} w^{\varepsilon}(k',x)dk'. \end{aligned} \tag{4.14}$$

If we put

$$y = \frac{x}{\varepsilon^2}, \tag{4.15}$$

$$L_1 = k\cdot\nabla_y, \qquad L_3 = k\cdot\nabla_x, \tag{4.16}$$

$$L_2 = ic_n \int \hat{V}(2k-2k')e^{-2i(k-k')\cdot y}\cdot dk'$$

$$- ic_n \int \hat{V}^*(2k-2k')e^{2i(k-k')\cdot y}\cdot dk', \tag{4.17}$$

and let $v^\varepsilon(x,y,k)$ be a solution of

$$(L_1 + \varepsilon L_2 + \varepsilon^2 L_3)v^3 = 0, \tag{4.18}$$

then† $w^\varepsilon(k,x) = v^\varepsilon(x,x/\varepsilon^2,k)$ will be a solution of (4.14).

The usual second order perturbation theory on (4.18) now yields the following equation for the first approximation to $E\{w^\varepsilon(k,x)\}$ which we denote $\tilde{w}(k,x)$.

$$k\cdot\nabla_x\tilde{w}(k,x) = \tilde{c}_n \int_{|k'|=k} \hat{R}(k-k')\tilde{w}(k',x)dk'$$

$$- \left\{c_n \int_{|k'|=k} \hat{R}(k-k')dk'\right\}\tilde{w}(k,x). \tag{4.19}$$

Here, $\tilde{c}_n$ are constants (related to c_n) and $\hat{R}$ is the power spectrum of the inhomogenities $V(x)$. Note that (4.19) is a genuine transport equation.

The details of the formal perturbation argument that lead from (4.18) to (4.19) are routine. We make good use of

---

†This is the usual multiscaling process.

(4.13), i.e. the stationarity of $V(x)$. Of course the passage from (4.14) to (4.19) is the problem that eludes analysis. Consideration of boundary conditions for (4.19) from boundary conditions for (4.1) is an even more difficult problem. Any mathematical development towards solving these problems would be an important accomplishment.

Appendix A. Law of Large Numbers for Stochastic Differential Equations.

Let $y(t)$, $t \geq 0$, be an ergodic Markov process on a compact separable metric space S with paths in $D([0,T];S)$. Let $P(t,y,A)$ be its transition functions and assume it induces a strongly continuous semigroup on $C(S)$ with infinitesimal generator Q defined on $D_Q \subset C(S)$. Let $\overline{P}$ be the invariant measure of P.

Let $x^{\varepsilon}(t)$ be the R^n-valued process defined by

$$\frac{dx^{\varepsilon}(t)}{dt} = F(x^{\varepsilon}(t),y^{\varepsilon}(t)), \qquad x^{\varepsilon}(0) = x, \qquad (A.1)$$

where $F(x,y)$ is smooth in x and a bounded continuous function on $R^n \times S \to R^n$. Here we have defined

$$y^{\varepsilon}(t) = y(t/\varepsilon), \qquad (A.2)$$

so that the coefficients in the equation (A.2) are rapidly varying $(\varepsilon \ll 1)$ but equilibrate since $y(t)$ is ergodic. We wish to analyze the behavior of $x^{\varepsilon}(t)$, $0 \leq t \leq T < \infty$, as $\varepsilon \to 0$.

Assume that

$$\lim_{T\uparrow\infty} \frac{1}{T} \int_0^T dt \int P(t,y,dz)F(x,z) = \overline{F}(x), \qquad (A.3)$$

defines a bounded smooth function $\overline{F}$ on $R^n \to R^n$, with (A.3) taken uniformly in x and y. Let

$$G^{\lambda}(x,y) = \int_0^{\infty} dt\ e^{-\lambda t} \int P(t,y,dz)[F(x,z)-\overline{F}(x)], \quad \lambda > 0. \quad (A.4)$$

From (A.3) it follows that $\lambda G^{\lambda} \to 0$, $\lambda \to 0$, uniformly in (x,y). Assume that the functions† $G_i^{\lambda}G_j^{\lambda}$, $i,j = 1,2,\cdots,n$, are in the domain of Q and that $QG_i^{\lambda}G_j^{\lambda}$ is a uniformly bounded function of (x,y) for each $\lambda > 0$.

Let $\overline{x}(t) \in R^n$ be defined by

$$\frac{d\overline{x}(t)}{dt} = \overline{F}(\overline{x}(t)), \quad \overline{x}(0) = x. \qquad (A.5)$$

We have the following result.

THEOREM. Under the above hypotheses

$$\lim_{\varepsilon\downarrow 0} P_{x,y}\{\sup_{0\leqslant t\leqslant T} |x^{\varepsilon}(t) - \overline{x}(t)| > \delta\} = 0 \qquad (A.6)$$

for all $\delta > 0$, $T < \infty$ and $x \in R^n$, $y \in S$.

REMARK 1. If $F(x,y) = \chi_A(y)$ $(n = 1)$, $A \subset S$, then

$$x^{\varepsilon}(t) = x + \int_0^t \chi_A(y(s/\varepsilon))ds,$$

$$\overline{x}(t) = x + t\overline{P}(A). \qquad (A.7)$$

---

† G_i^{λ} is the ith component of G^{λ}.

Here, $\chi_A(y)$ is the characteristic function of the set A. Note that in this case (A.6) is a well-known fact, i.e. the fraction of time spent by $y(t)$ in A tends in probability to $\overline{P}(A)$ as $t \uparrow \infty$.

PROOF. Clearly $(x^\varepsilon(t), y^\varepsilon(t))$ constitute a Markov process on $R^n \times S$ with generator

$$\frac{1}{\varepsilon} Q + F \cdot \frac{\partial}{\partial x}. \tag{A.8}$$

From (A.4) it follows that

$$\begin{aligned} x^\varepsilon(t) + \varepsilon G^\lambda(x^\varepsilon(t), y^\varepsilon(t)) &= x + \varepsilon G^\lambda(x,y) + \int_0^t \overline{F}(x^\varepsilon(s))ds \\ &+ \int_0^t \left[\lambda G^\lambda(x^\varepsilon(s), y^\varepsilon(s)) + \varepsilon F(x^\varepsilon(s), y^\varepsilon(s)) \cdot \frac{\partial G^\lambda(x^\varepsilon(s), y^\varepsilon(s))}{\partial x}\right] ds \\ &+ M^{\varepsilon,\lambda}(t), \end{aligned} \tag{A.9}$$

where $M^{\varepsilon,\lambda}(t) = (M_i^{\varepsilon,\lambda}(t))$ is an integrable right-continuous martingale (vector-valued) with mean zero. By direct computation we find that the increasing processes associated with $M^{\varepsilon,\lambda}(t)$ are given by

$$\begin{aligned} \langle M_i^{\varepsilon,\lambda}(t), M_j^{\varepsilon,\lambda}(t)\rangle = \varepsilon \int_0^t [QG_i^\lambda G_j^\lambda(x^\varepsilon(s), y^\varepsilon(s)) - G_i^\lambda Q G_j^\lambda(x^\varepsilon(s), y^\varepsilon(s)) \\ - G_j^\lambda Q G_i^\lambda(x^\varepsilon(s), y^\varepsilon(s))]ds. \end{aligned} \tag{A.10}$$

The integrands on the right side of (A.10) are uniformly bounde functions of x and y for all $\lambda > 0$, by hypothesis.

Since $\overline{F}(x)$ has bounded derivatives by hypotheses, (A.5) and (A.9) imply that

$$\sup_{0\leq t\leq T} |x^{\varepsilon}(t) - \overline{x}(t)| \leq C \sup_{0\leq t\leq T} |g^{\varepsilon,\lambda}(t)|, \quad \text{(A.11)}$$

where C is a constant, $|x|^2 = \sum_{i=1}^{n} X_i^2$ and

$$g^{\varepsilon,\lambda}(t) = \varepsilon[G^{\lambda}(x,y) - G^{\lambda}(x^{\varepsilon}(t),y^{\varepsilon}(t))]$$

$$+ \int_0^t \left[\lambda G^{\lambda}(x^{\varepsilon}(s),y^{\varepsilon}(s)) + \varepsilon F(x^{\varepsilon}(s),y^{\varepsilon}(s)\cdot\frac{\partial G^{\lambda}(x^{\varepsilon}(s),y^{\varepsilon}(s)}{\partial x}\right]ds$$

$$+ M^{\varepsilon,\lambda}(t). \quad \text{(A.12)}$$

Since $\lambda G^{\lambda} \to 0$ as $\lambda \to 0$ uniformly in (x,y) and $F\cdot\partial G^{\lambda}/\partial x$ is bounded, the first two terms on the right side of (A.12) are small, if λ is chosen small and then ε is chosen small. Hence, it suffices to show that

$$\lim_{\varepsilon\downarrow 0} P_{x,y}\{\sup_{0\leq t\leq T} |M^{\varepsilon,\lambda}(t)| > \delta\} = 0,$$

$$\lambda > 0, \quad \delta > 0. \quad \text{(A.13)}$$

However, $|M^{\varepsilon,\lambda}(t)|$ is a submartingale and by Kolmogorov's inequality and (A.10), we have $(\lambda > 0)$

$$P_{x,y}\{\sup_{0\le t\le T}|M^{\varepsilon,\lambda}(t)| > \delta\} \le \frac{E\{|M^{\varepsilon,\lambda}(T)|^2\}}{\delta^2}$$

$$= \frac{1}{\delta^2}\sum_{i=1}^{n} \langle M_i^{\varepsilon,\lambda}(T), M_i^{\varepsilon,\lambda}(T)\rangle$$

$$\le \frac{C'}{\delta^2}\,\varepsilon,$$

where C' is a constant. This proves the Theorem.

REMARK 2. The method of averaging for deterministic ODE's is a special case of the theorem.

REMARK 3. Suppose that $\overline{F}(0) = 0$ and that $x = 0$ is an asymptotically stable equilibrium point for (A.5). That is, if $|x(0)| < \delta$,

$$\lim_{t\uparrow\infty} |\overline{x}(t)| = 0. \tag{A.14}$$

Suppose that $F(0,y) = 0$ also. Then one can conclude that, given u_1 and u_2 positive, there is a δ_1 and an ε_0 such that for ε fixed, $0 < \varepsilon \le \varepsilon_0$, $|x^{\varepsilon}(0)| < \delta_1$,

$$P_{x,y}\{\sup_{t\ge 0} |x^{\varepsilon}(t)| > u_1\} < u_2. \tag{A.15}$$

Results of this form are analyzed systematically in a forthcoming paper by Blankenship and Papanicolaou.

APPENDIX B. GAUSS MARKOV LIMIT THEOREM FOR STOCHASTIC DIFFERENTIAL EQUATIONS. The theorem of the previous section tells us that $x^{\varepsilon}(t)$ of (A.1) and $\overline{x}(t)$ of (A.5) are close as $\varepsilon \downarrow 0$. Define the *fluctuation process*

$z^{\varepsilon}(t)$ by

$$x^{\varepsilon}(t) = \overline{x}(t) + \sqrt{\varepsilon}\, z^{\varepsilon}(t). \tag{B.1}$$

The purpose of this appendix is to find the limit law of $z^{\varepsilon}(t)$, $0 \leqslant t \leqslant T < \infty$, as $\varepsilon \downarrow 0$ (cf. [13], [34] and Appendix A of I).

Let us assume for simplicity that Q, the infinitesimal generator of $y(t)$, $t \geqslant 0$, is a bounded operator on $C(S)$ and that the recurrent potential kernel

$$\psi(y,A) = \int_0^{\infty} dt[P(t,y,A) - \overline{P}(A)], \quad y \in S, \quad A \subset S, \tag{B.2}$$

defines a bounded operator on $C(S)$.

Define $A_{ij}(x)$, $i,j = 1,2,\cdots,n$ by

$$A_{ij}(x) = \iint_{SS} \overline{P}(dy)\psi(y,dz)(F_i(x,y) - \overline{F}_i(x))(F_j(x,z) - \overline{F}_j(x)). \tag{B.3}$$

We have the following.

THEOREM. With the hypotheses of Appendix A and the ones above, the process $(z^{\varepsilon}(t),\overline{x}(t))$ on $R^n \times R^n$, $0 \leqslant t \leqslant T$, converges weakly as $\varepsilon \to 0$ (as a measure on $C([0,T];\, R^n \times R^n)$) to the diffusion Markov process $(z(t),\overline{x}(t))$ with generator

$$f(z,\overline{x}) = \sum_{i,j=1}^{n} A_{ij}(\overline{x}) \frac{\partial}{\partial z_i \partial z_j} + \sum_{i,j=1}^{n} \frac{\partial \overline{F}_i(\overline{x})}{\partial \overline{x}_j} z_j \frac{\partial}{\partial z_i}$$

$$+ \sum_{i=1}^{n} \overline{F}_i(\overline{x}) \frac{\partial}{\partial \overline{x}_i}. \tag{B.4}$$

REMARK. Clearly $z(t)$, the limiting fluctuation process, is a Gaussian process with independent increments. Let $U(t,s)$ be the fundamental solution of the variational equation associated with (A.5), i.e.

$$\frac{dU(t,s)}{dt} = \frac{\partial \bar{F}(\bar{x}(t))}{\partial \bar{x}} U(t,s), \quad t > s, \quad U(s,s) = I. \tag{B.5}$$

If $z(0) = 0$, as is natural from (B.1), it follows that

$$E\{z(t)\} = 0, \tag{B.6}$$

and†

$$E\{z(t)z^T(t)\} = \int_0^t U(t,s)A(\bar{x}(s))U^T(t,s)ds. \tag{B.7}$$

Thus, the statistical properties of the limiting fluctuation process are easy to describe. One does not have to solve a Fokker-Planck equation (a PDE). Of course the result itself is a relatively crude one compared with the full diffusion approximations (cf. Appendix A of Part I and the many references cited there).

PROOF. We consider $(z^\varepsilon(t), \bar{x}(t), y^\varepsilon(t))$ jointly which is a Markov process on $R^n \times R^n \times S$ with generator

$$L^\varepsilon = \frac{1}{\varepsilon} Q + \frac{1}{\sqrt{\varepsilon}} (F(\bar{x} + \sqrt{\varepsilon} z, y) - \bar{F}(\bar{x}))\cdot\frac{\partial}{\partial z} + \bar{F}(\bar{x})\cdot\frac{\partial}{\partial x}. \tag{B.8}$$

---

† z^T is the transpose of the column vector z.

Of course $\bar{x}(t)$ is deterministic, independently of $x^{\varepsilon}(t)$, $y^{\varepsilon}(t)$, etc., but it is kept along to make things time homogeneous.

It is easy to see that since F and $\bar{F}$ are bounded (along with x-derivatives), moments of $z^{\varepsilon}(t)$ are uniformly bounded on bounded t intervals (uniformly in ε also). We are therefore working, effectively, on a compact state space for $(x^{\varepsilon},\bar{x},y^{\varepsilon})$.

The proof that $(z^{\varepsilon},\bar{x})$ converges weakly to (z,x), associated with $\bar{L}$ in (B.4), goes in two steps. In Step 1 we show that $(z^{\varepsilon},\bar{x})$ are relatively weakly compact in $D([0,T];\ R^n \times R^n)$. In Step 2 we identify the limit process and this is elementary since no special theory is needed for $\bar{L}$; it has, essentially, constant coefficients.

Both steps use the following preliminary steps.

Let $f(z,\bar{x})$ be a smooth real function and define f_1 and f_2 by

$$f_1(z,\bar{x},y) = \int \psi(y,d\zeta)\Big[[F(\bar{x},\zeta) - \bar{F}(\bar{x})]\cdot\frac{\partial f_1(z,\bar{x},\zeta)}{\partial z} + \frac{\partial F(\bar{x},y)}{\partial \bar{x}}\, z\cdot\frac{\partial f(z,\bar{x})}{\partial z} + \bar{F}(\bar{x})\cdot\frac{\partial f(z,\bar{x})}{\partial \bar{x}} - \bar{L}f(z,\bar{x})\Big]. \qquad (B.9)$$

Then,

$$Qf_1 + (F(\bar{x},y) - \bar{F}(\bar{x}))\cdot\frac{\partial f(z,\bar{x})}{\partial z} = 0 \tag{B.10}$$

$$Qf_2 + (F(\bar{x},y) - \bar{F}(\bar{x}))\cdot\frac{\partial f_1(z,\bar{x},y)}{\partial z} + \frac{\partial F(\bar{x},y)}{\partial \bar{x}}\, z\cdot\frac{\partial f(z,\bar{x})}{\partial z}$$
$$+ \bar{F}(\bar{x})\cdot\frac{\partial f(z,\bar{x})}{\partial \bar{x}} - \bar{L}f = 0. \tag{B.11}$$

If we let

$$f^{\varepsilon}(z,\bar{x},y) = f(z,\bar{x}) + \sqrt{\varepsilon}\, f_1(z,\bar{x},y) + \varepsilon f_2(z,\bar{x},y), \tag{B.12}$$

then, it follows from (B.8)-(B.12) that

$$L^{\varepsilon}f^{\varepsilon} = \bar{L}f + g^{\varepsilon} = \bar{L}f + O(\sqrt{\varepsilon}). \tag{B.13}$$

Here $O(\sqrt{\varepsilon})$ contains powers of z, but is otherwise uniform. Since $z^{\varepsilon}(t)$ moments are bounded, as we pointed out earlier, the $O(\sqrt{\varepsilon})$ in (B.13) is effectively uniform.

The above constructions and the definition of the martingale

$$M_{f^{\varepsilon}}(t) = f^{\varepsilon}(z^{\varepsilon}(t),\bar{x}(t),y^{\varepsilon}(t)) - f^{\varepsilon}(z,\bar{x},y)$$
$$+ \int_0^t L^{\varepsilon}f^{\varepsilon}(z^{\varepsilon}(s),\bar{x}(s),y^{\varepsilon}(s))ds \tag{B.14}$$

lead to the following identity.

$$f(z^{\varepsilon}(t),\bar{x}(t)) = f(z,\bar{x}) + \varepsilon[f_1(z,\bar{x},y) - f_1(z^{\varepsilon}(t),\bar{x}(t),y^{\varepsilon}(t))$$
$$+ \varepsilon f_2(z,\bar{x},y) - \varepsilon f_2(z^{\varepsilon}(t),\bar{x}(t),y^{\varepsilon}(t))] + \int_0^t \bar{L}f(z^{\varepsilon}(s),\bar{x}(s))ds$$

$$+ \int_0^t g^\varepsilon(z^\varepsilon(s),\bar{x}(s),y^\varepsilon(s))ds + M_{f^\varepsilon}(t). \tag{B.15}$$

The martingale $M_{f^\varepsilon}(t)$ is right continuous and integrable (since $z^\varepsilon(t)$ has moments) and its increasing process is given by

$$\langle M_{f^\varepsilon}(t),M_{f^\varepsilon}(t)\rangle = \int_0^t [Q(f_1 + \varepsilon f_2)^2 - 2(f_1 + \varepsilon f_2)Q(f_1 + \varepsilon f_2)]ds, \tag{B.16}$$

as can be verified by direct computation.

From (B.15) and (B.16), one can immediately deduce compactness in D by choosing f to be coordinate functions. Thus, Step 1 is complete.

For Step 2 the task is almost complete in view of (B.15) because (F_t is the σ-algebra generated by the paths up to time t)

$$\lim_{\varepsilon \downarrow 0} E\left\{ |E\{f(z^\varepsilon(t),\bar{x}(t)) - f(z^\varepsilon(s),\bar{x}(s)) - \int_s^t \bar{L}f(z^\varepsilon(\sigma),\bar{x}(\sigma))d\sigma | F_s\}| \right\} = 0. \tag{B.17}$$

The result (B.17) implies immediately that the limit process is uniquely identified with the one generated by $\bar{F}$ of (B.4).

We have omitted a few details, of course, but the basic ideas should be clear; the details can be provided easily.

Virginia Polytechnic Institute and State University.
Courant Institute of Mathematical Sciences, New York University.

RELATIONS BETWEEN SAMPLE AND MOMENT STABILITY FOR LINEAR STOCHASTIC DIFFERENTIAL EQUATIONS

BY

F. KOZIN AND S. SUGIMOTO

1. INTRODUCTION.

A problem that has been of interest almost since the beginning of studies of stability of stochastic systems is the relationships and implications that exist between the various stability definitions that have been suggested and studied for these systems.

Students of these questions have been concerned with stability in probability, stability of the moments and almost sure stability of the sample solutions themselves [1].

It is known that the sample solutions may be stable with probability one for some region of the system parameters, and yet in that same region the second moments may grow exponentially. Or, in some region the second moments decay exponentially (stability in the mean square sense) yet in the same region the fourth moments will grow exponentially.

This combination of known facts has often led to question just how practically useful any particular stability concept is for stochastic systems and what in fact do any of them mean.

We have generally stated, in previous manuscripts, that only sample stability can be meaningful for applications of stochastic models to real phenomena. Yet even though a stochastic system may possess stable sample solutions, the solutions may possess other characteristics, such as exceeding critical values or decaying very slowly, that make this property impractical without further understanding of its implications.

In this paper we will attempt to shed some light upon the problem by establishing an exact relation between the region of sample stability and regions of moment stability for linear Ito stochastic differential equations. This establishes a new characterization of sample stability regions which can be useful in simulation studies. Some of these ideas can be extended to the physical noise coefficient case but in a less rigorous fashion. We shall first motivate our results by the first order linear Ito equation.

2. A Simple Illustrative Example.

We shall motivate our results by studying the simple first order stochastic differential equation with a white noise coefficient.

We consider the first order Ito equation

$$dx_t + (adt + dB_t)x_t = 0, \tag{2.1}$$

where a is constant and B_t represents the Brownian Motion process with zero mean and variance $\sigma^2 t$.

It is well known that the solution process of (2.1) is

$$x_t = x_0 e^{-(a+\frac{\sigma^2}{2})t - B_t}. \tag{2.2}$$

Again, it is well known that since B_t grows like $\sqrt{t \log \log t}$ with probability one, then the stability properties of the sample solutions are determined by the deterministic term in the exponent of (2.2). Hence, the region of sample stability is clearly given by $a + \sigma^2/2 > 0$, i.e.

$$\sigma^2 > -2a. \tag{2.3}$$

We now consider the moments of the solution process (2.2). The absolute p^{th} moment is simply found to be

$$\begin{aligned} E\{|x(t)|^p\} &= |x_0|^p \, e^{-p(a+\frac{\sigma^2}{2})t} E\{e^{-pB(t)}\} \\ &= |x_0|^p \, e^{\frac{pt}{2}[(p-1)\sigma^2 - 2a]}, \end{aligned} \tag{2.4}$$

where $x(0) = x_0$ with probability one.

The region of stability for the integer moment $p = n(>1)$ is given by (2.4) as

$$\sigma^2 < \frac{2}{(n-1)} a. \tag{2.5}$$

From (2.5) it is obvious that the constant a must be greater than zero for stability of the nth moment. This is not necessary for sample stability as given by the region (2.3). As n increases the stability region decreases and lies below the line $\sigma^2 = \frac{2}{n-1} a$, $(a > 0)$ in the (a,σ^2) plane.

This illustrates the known facts about stability for the first order Ito differential equation. The regions of stability for higher integer moments are included in the regions of stability of lower integer moments and all integer moment stability regions are included in the region of sample stability.

Going back to (2.4), we notice that the equality holds for all $p \geqslant 0$. Therefore, let us break from previous studies of integer moments and consider the case $0 < p < 1$. For this range of values of p, the stability region obtained from (2.4) is

$$\sigma^2 > \frac{2}{p-1} a, \quad (0 < p < 1). \tag{2.6}$$

A rather curious and interesting result follows from (2.6). We see that as p approaches zero, the half line that bounds the stability region for the pth moment (the other boundary is the positive a-axis) rotates counterclockwise and approaches the half line that bounds the region of sample stability. It appears, at least for the simple stochastic equation (2.1), that the region of sample stability can be characterized as the union of all p^{th} moment stability regions for $p > 0$. The next obvious question is, does this property hold more generally?

Similar results can be established for the first order stochastic differential equation

$$\frac{dx(t)}{dt} + [a + f(t)]x(t) = 0, \quad x(0) = x_0, \tag{2.7}$$

where the f-process is stationary, ergodic and Gaussian with zero mean. The sample stability region is given simply by $a > 0$.

The pth moment stability region can be shown to be given by the inequality

$$a > \frac{p}{2}s(0), \quad \text{for} \quad p > 0, \tag{2.8}$$

where $s(\omega)$ is the spectral density of the f-process.

Clearly as in the Ito case above, as p approaches zero, the pth moment stability regions approach the sample stability region.

In each of the two examples above, we see that the regions of sample stability are determined by the system parameter values that satisfy the inequality

$$\lim_{p \downarrow 0} \frac{1}{pt} \log E\{|x(t)|^p / \|x_0\|^p\} < 0 \tag{2.9}$$

for all $t > 0$.

In the next section, we shall see that it is the limit (2.9) that plays the basic role for determining the sample stability region for higher order linear Ito stochastic differential equations.

3. MAIN RESULTS. The basic objective of this section is to present a result that establishes a limit similar to (2.9), which characterizes the region of sample stability for arbitrary linear homogeneous Ito stochastic differential equations. The proof of the result follows the technique of Littlewood [2], and requires results of Dynkin [3], as well as Khasminskii [4].

We first establish preliminary lemmas [5]. For these lemmas, we require the following notations. Let x denote the solution vector of the stochastic differential equation.

$$dx(t) = Fx(t)dt + \sum_{r=1}^{m} G^r x(t) dB^r(t), \quad (3.1)$$

where x is an n-vector, F is an $n \times n$ constant matrix, G^r, $r = 1,\cdots,m$ are also $n \times n$ constant matrices, and $B^r(t)$ $r = 1,\cdots,m$ are m independent Brownian motion processes with zero means and variances $\sigma_r^2(t) = t$, $r = 1,\cdots,m$.

Analogously to Littlewood's deterministic development, for the random vector x, we define the quantities

$$\begin{aligned} M_p(x) &= E\{\|x\|^p\}^{1/p}, \\ L_p(x) &= \log M_p(x), \\ M_o(x) &= \exp L_o(x), \\ L_o(x) &= E\{\log \|x\|\}, \end{aligned} \quad (3.2)$$

where

$$\|x\| = \left\{\Sigma x_i^2\right\}^{1/2},$$

which are studied in the following lemmas.

LEMMA 3.1. For $0 < p_1 \leqslant p_2$, if $M_{p_1}(x)$, $M_{p_2}(x)$ exist, then $M_{p_1}(x) \leqslant M_{p_2}(x)$.

PROOF. Let $f(x) = x^{p_2/p_1}$, which is convex since $p_2/p_1 \geqslant 1$. Thus, by a simple application of Jensen's theorem, for $f(x)$, we have

$$E\left\{\|x\|^{p_1}\right\}^{p_2/p_1} \leqslant E\left\{\|x\|^{p_2}\right\}$$

from which the result follows.

LEMMA 3.2. Assume M_p exists for some $p > 0$. Then there exists M_{0+} for which

$$\lim_{p \downarrow 0} M_p = M_{0+}.$$

PROOF. The proof is trivial since by lemma 3.1, M_p is monotone decreasing as p approaches zero from above and $M_p \geqslant 0$.

LEMMA 3.3. Assume M_p exists for some positive p. Then the following limits exist and the equality holds.

$$L_{0+} = \lim_{p \downarrow 0} L_p = \lim_{p \downarrow 0} \log E\left\{\|x\|^p\right\}^{1/p} = E\left\{\log \|x\|\right\} = L_0. \quad (3.3)$$

PROOF. We shall establish the inequalities $L_{0+} \geqslant L_0$

and $L_o \geqslant L_{o+}$.

(i) $L_{o+} \geqslant L_o$. Clearly, for $p > 0$,

$$\|x\|^p = \exp(p \log\|x\|) \geqslant 1 + p \log\|x\|.$$

Thus,

$$E\left\{\|x\|^p\right\} \geqslant 1 + p\, E\left\{\log\|x\|\right\}. \tag{3.4}$$

Raising each side of (3.4) to $1/p$ and taking limits yields from lemma 3.2, and (3.2)

$$\begin{aligned} M_{o+} = \left(E\left\{\|x\|^p\right\}\right)^{1/p} &\geqslant \lim_{p\downarrow 0}\left(1 + p\, E\left\{\log\|x\|\right\}\right)^{1/p} \\ &= \lim_{p\downarrow 0}(1 + p\, L_o)^{1/p} \\ &= e^{L_o} \\ &= M_o. \end{aligned} \tag{3.5}$$

Hence, $M_{o+} \geqslant M_o$, and by the monotone property of the logarithm yields,

$$L_{o+} = \log M_{o+} \geqslant \log M_o = L_o.$$

(ii) $L_{o+} \leqslant L_o$. We shall first assume that $\|x\| \geqslant 1$, for which it follows that $\log\|x\| \geqslant 0$.

By Taylor's theorem, we have

$$\|x\|^p = \exp(\log\|x\|^p) \leqslant 1 + p \log\|x\| + (p \log\|x\|)^2(1 + \|x\|^{\theta p}),$$

$$0 < \theta < 1. \tag{3.6}$$

We take expectations of (3.6), using the notation of (3.2) to obtain,

$$M_p^p(x) \leq 1 + p\, L_o(x) + p^2 K(x), \qquad (3.7)$$

where

$$K = E\left\{(\log\|x\|)^2(1 + \|x\|^{\theta p})\right\}$$

$$\leq \left|E\left\{(\log\|x\|)^4\right\}E\left\{(1 + \|x\|^{\theta p})^2\right\}\right|^{1/2}$$

$$\leq \left|E\left\{(\log\|x\|)^4\right\}\left\{1 + E\left\{\|x\|^{\theta p}\right\}\right\}\right|^{1/2}$$

$$< \infty,$$

since

$$E\left\{\|x\|^{\theta p}\right\} \leq M_p^{\theta p} < \infty,$$

and

$$E\left\{(\log\|x\|)^4\right\} < \infty$$

which follows from

$$\|x\|^p \geq \left(\frac{p}{4}\right)^4 (\log\|x\|)^4.$$

Thus, from the boundedness of K, (3.7) yields

$$M_{o+} = \lim_{p\downarrow o} M_p \leq \lim_{p\downarrow o}(1 + pL_o + p^2K)^{1/p} = e^{L_o},$$

or by taking logarithms of each side,

$$L_{o+} = \lim_{p \downarrow o} L_p \leq L_o. \tag{3.8}$$

Now, on the basis of the $\log\|x\| > 0$ case just considered, we can define

$$x_n = \begin{cases} x, & \|x\| > n^{-1} \\ n^{-1}, & \|x\| \leq n^{-1}, \end{cases}$$

which implies,

$$\|x_n\| = \max(\|x\|, n^{-1}).$$

Clearly, $M_p(x) \leq M_p(x_n)$, thus, this inequality holds for the logarithms as well; hence, we have, by (3.8) for the $\log\|x\| > 0$ case,

$$L_{o+}(x) = \lim_{p \downarrow o} L_p(x) \leq \lim_{p \downarrow o} L_p(x_n) \leq L_o(x_n),$$

for any n. Thus, we find

$$L_{o+}(x) \leq \lim_{n \uparrow \infty} L_o(x_n) = L_o(x),$$

since, for $E_n = \{\|x\| < n^{-1}\}$, we have

$$\left|\int_{E_n} \log\|x_n\| dP\right| \leq \left|\int_{E_n} \log\|x\| dP\right|$$

and

$$\lim_{n \uparrow \infty} \left|\int_{E_n} \log\|x\| dP\right| = 0.$$

Hence, our lemma is proved.

We have just established the equality

$$\lim_{p \downarrow o} \frac{1}{p} \log E\left\{e^{py}\right\} = E\left\{y\right\}, \qquad (3.9)$$

for a random variable y, assuming that there exists a $p_1 > 0$ such that

$$E\left\{e^{p_1 y}\right\} < \infty.$$

In order to apply these results to our problem, we now return to our original stochastic differential equation (3.1). Upon rewriting this equation in component form as,

$$dx_i(t) = \sum_{j=1}^{m} F_{ij}x_j(t)dt + \sum_{r=1}^{m}\sum_{j=1}^{m} G^r_{ij}x_j(t)dB^r(t), \quad i = 1,\cdots,n, \qquad (3.10)$$

we can introduce the logarithm, $\log\|x(t)\|$, and apply Ito's differential rule to obtain

$$d\log\|x(t)\| = \mathcal{L}\log\|x(t)\|dt + \sum_{r=1}^{m}\sum_{i=1}^{m} \frac{\partial \log\|x(t)\|}{\partial x_i} G^r_i x(t)dB^r(t), \qquad (3.11)$$

where G^r_i denotes the ith row vector of the matrix G^r and $\mathcal{L}$ is the backward operator generated by (3.10). That is,

$$\mathcal{L} = \left(\frac{\partial}{\partial x}, Fx\right) + \frac{1}{2}\sum_{i,j=1}^{m} \sigma_{ij}(x)\frac{\partial^2}{\partial x_i \partial x_j},$$

where

$$\sigma_{ij}(x) = \sum_{r=1}^{m}\sum_{k,\ell=1}^{m} G^r_{ik}G^r_{j\ell}x_k x_\ell.$$

If one defines the unit vector

$$\lambda(t) = x(t)/\|x(t)\|, \quad \|\lambda(t)\| \equiv 1,$$

then the drift and diffusion terms of (3.11) can be written in terms of the λ vector as

$$\mathcal{L} \log\|x\| = \left[(F\lambda,\lambda) + \frac{1}{2}\sum_{i=1}^{m} a_{ii}(\lambda) - \sum_{i,j=1}^{m} a_{ij}(\lambda)\lambda_i\lambda_j\right] \equiv Q(\lambda), \tag{3.12}$$

where

$$a_{ij}(\lambda) = \sum_{k,s=1}^{m} \sum_{r=1}^{m} G^r_{ik} G^r_{js} \lambda_k \lambda_s, \quad G^r = (G^r_{ij}),$$

and

$$\sum_{r=1}^{m} \sum_{i=1}^{m} \frac{\partial \log\|x\|}{\partial x_i} G^r_i x \, dB^r = \sum_{r=1}^{m} (G^r\lambda,\lambda) dB^r \equiv \sum_{r=1}^{m} R^r(\lambda) dB^r. \tag{3.13}$$

We note that the functions of λ that appear in (3.12) and (3.13) are polynomials and are bounded since the λ-vector is a unit vector, which is therefore constrained to lie on the surface of the unit n-sphere. The λ-process may be ergodic on the entire surface of the n-sphere or it may possess ergodic components which are subsets of the surface. It can be shown that there are no more than n-distinct ergodic components on the surface of the n-dimensional sphere [4]. For a complete discussion of the ergodic components for the two dimensional λ-vector on the boundary of the unit circle, see [6].

It is also known that the limit of the Ito integral,

$$\lim_{t\uparrow\infty} \frac{1}{t} \int_0^t h(y(t))dB(t) = 0 \tag{3.14}$$

holds almost surely for the non-anticipating y-process as long as $E\{h^2(y(t))\}$ is bounded [4]. Now, upon integrating (3.11) and using the notation of (3.12), (3.13), we obtain

$$\log\|x(t)\| - \log\|x_0\| = \int_0^t Q(\lambda(\tau))d\tau + \sum_{r=1}^{m} R^r(\lambda(\tau))dB^r(\tau). \tag{3.15}$$

Thus, if the λ process is ergodic on the entire sphere, we can divide (3.15) by t and let t approach infinity. Since $R^r(\lambda)$ is bounded, then (3.14) will hold for $R^r(\lambda(t))$, $r = 1,\cdots,m$, and we obtain

$$\lim_{t\uparrow\infty} \frac{\log\|x(t)\| - \log\|x_0\|}{t} = \lim_{t\uparrow\infty} \frac{1}{t} \int_0^t Q(\lambda(\tau))d\tau = E\{Q(\lambda(\tau))\} \tag{3.16}$$

almost surely, where the expectation is taken over the invariant λ-measure.

As first pointed out by Khasminskii, (3.16) establishes that $E\{Q(\lambda(\tau))\} < 0$ is the necessary and sufficient condition for almost sure sample stability. It is this last condition that allows us to establish the connection between sample stability and moment stability. Equation (3.15) is equivalent to the exponential equality,

$$\|x(t)\| = \|x_0\| e^{\left[\int_0^t Q(\lambda(\tau))d\tau + \sum_{r=1}^{m} \int_0^t R^r(\lambda(\tau))dB^r(\tau) \right]} . \quad (3.17)$$

Since, Q and R^r are bounded functions, an application of the following theorem [3, thm. 7.3, p. 234] to (3.17) establishes the existence of the expected value, for each t, of $\|x(t)\|$.

THEOREM (DYNKIN). Let $0 \leqslant s \leqslant t \leqslant \infty$. Assume that, for any $u \in (s,t)$ the functions $V(u)$ and $\Phi(u)$ satisfy the inequality

$$V(u) + \frac{1}{2} \Phi^2(u) \leqslant 0,$$

almost surely. Then

$$E\left\{ e^{\int_s^t V(u)du + \int_s^t \Phi(u)dB(u)} \right\} \leqslant 1.$$

Further, assume that for some constant c, $|V(u)| \leqslant c$ almost surely for all $u \in (s,t)$. Then, for any real numbers λ_1, λ_2,

$$E\left\{ e^{\lambda_1 \int_s^t V(u)du + \lambda_2 \int_s^t \Phi(u)dB(u)} \right\} \leqslant e^{|\lambda_1 - \lambda_2^2| c(\lambda - s)} .$$

Hence, $E\{\|x(t)\|\}$ is bounded for each t, which allows us to apply lemma 3.3 for each t in $(0,\infty)$, to (3.17), obtaining

$$\lim_{p\downarrow o} \frac{1}{p} \log E\{\|x(t)\|^p/\|x_o\|^p\} = E\{\log\frac{\|x(t)\|}{\|x_o\|}\}$$

$$= E\{\int_o^t Q(\lambda(\tau))d\tau\}$$

$$+ \sum_{r=1}^{m} E\{\int_o^t R^r(\lambda(\tau))dB^r(\tau)\} \tag{3.18}$$

$$= t\ E\{Q(\lambda)\},$$

where the expectation on the right is with respect to the invariant λ-measure.

Since $E\{Q(\lambda)\} < 0$ defines the region of sample stability for the case that λ is ergodic on the entire surface of the n-sphere, (3.18) establishes the characterization of the region of almost sure sample stability which can be considered to be the limit of the regions of stability of the moments

$$E\{\|x(t)\|^p\} = E\{[x_1^2(t) + \cdots + x_n^2(t)]^{p/2}\} \tag{3.19}$$

as p approaches zero.

Furthermore, since the inequalities

$$E\{\|x\|_p^p\} \geq E\{\|x\|^p\} \geq k\ E\{\|x\|_p^p\} \geq k\ E\{\|x\|^p\} \tag{3.20}$$

hold for $\|x\|_p^p = \sum_{i=1}^{m} |x_i|^p$, where $0 < p \leq 2$ and $k = m^{p/2} = 1$, it follows that the region of sample stability is the limit of the regions of pth moment stability for p approaching zero. Thus, we have established the desired connection between moment stability and sample stability for linear Ito stochastic differential equations, when λ is ergodic on the entire sphere.

4. CONCLUSION. In the previous section we established an exact relation between conditions for sample stability and conditions for moment stability of linear Ito differential equations. Computer simulation results for first and second order linear equations show that it may be possible to use this type of result to establish sample stability boundaries, although much computation will be needed for higher order systems. Furthermore, the fact that the structure of the ergodic components on the surface of the n-sphere is apparently unknown at this time complicates the situation significantly for higher order systems.

On the other hand, we now know that sample stability criteria will include samples that are stable in some p^{th} moment, no matter how small p may be. This implies that for certain regions the sample solutions may decay very slowly (especially for small p) which certainly would not be useful from the point of view of applications to the design of stable systems.

Thus, it may be that in order to specify a certain rate of decay of the samples, higher order moment stability constraints will be required. This brings forth various questions of relating sample decay to moment decay for higher order linear equations. It also brings up questions of relating uniform bounds to moment decay properties.

These are questions of importance in applications of stochastic differential equations. Answers to these questions would be quite welcome.

REFERENCES

1. F. Kozin, *A Survey of Stability of Stochastic Systems* , Automatica Vol. 5, Pergomon Press (1969), 95.

2. J. E. Littlewood, *Lectures on the Theory of Functions*, Oxford Univ. Press, London (1944).

3. E. B. Dynkin, *Markov Processes*, Academic, New York, Springer, Berlin (1965).

4. R. Z. Khasminskii, *Stability of Systems of Differential Equations with Random Parametric Excitation*, Izdat. Nauka, Moscow (1969).

5. S. Sugimoto, *Relations Between Sample and Moment Stability and Related Topics*, Ph.D. Thesis, Dept. of Electrical Engineering, Polytechnic Institute of New York (June 1974).

6. R. R. Mitchell and F. Kozin, *Sample Stability of Second Order Linear Differential Equations with Wide Band Noise Coefficients*, SIAM Journ. Appl. Math. 27 (1974), 571-605.

Polytechnic Institute of New York.
Osaka University.

DIFFUSION APPROXIMATIONS IN POPULATION GENETICS - HOW GOOD ARE THEY?

BY

BENNY LEVIKSON

1. INTRODUCTION. Diffusion approximations are widely used in applied probability. The reason is that these approximations provide convenient computational methods for approximating many important distributions of random processes and certain quantities associated with them such as stationary distributions, time to absorption, hitting probabilities and moments of stochastic integrals.

In the second section we show how to derive limiting diffusions for discrete Markov processes and how to study properties of these diffusions. In the third section we apply these methods to study several Markov processes arising in population biology by finding their limiting diffusions, analyzing these diffusions and inferring from the latter to the original process. Moreover, we show how to improve the accuracy of the results obtained from the diffusion approximations and we compare these results with some exact numerical computations and bounds.

In this paper we give what we consider to be the main ideas and methods for dealing with some problems in this field, while references for complete details are given throughout.

2. From Markov Chains to Diffusion.

To get a limiting diffusion for a Markov chain we rescale the state space and the time so that transitions occur at

$$t = nh \ (n = 0,1,2,\cdots)$$

and the rescaled process X_n^h satisfies

$$E\{\Delta X_n^h | X_n^h = x\} = h\mu(x, nh) + o(h),$$

$$E\{\Delta^2 X_n^h | X_n^h = x\} = h\sigma^2(x, nh) + o(h),$$

$$E\{|\Delta X_n^h|^\alpha | X_n^h = x\} = O(h), \qquad \alpha > 2.$$

Then the family $\{X_n^h\}$ is tight and

$$\Pr\{X_n^h \in B | X_0^h = x\} \to \Pr\{X(t) \in B | X(0) = x\} \text{ as } nh \to t,$$

where $X(t)$ is the unique diffusion with drift $\mu(x,t)$ and diffusion coefficient $\sigma^2(x,t)$. Furthermore, the multidimensional distribution of $\{X_{n_1}^h, X_{n_2}^h, \cdots, X_{n_k}^h\}$ converges to that of $\{X(t_1),\cdots,X(t_k)\}$, where $n_i h \to t_i$ $\{i = 1,2,\cdots,k\}$. Hence X_n^h converges weakly to $X(\cdot)$. The weak convergence assures that continuous functionals of the discrete process converge to those of the diffusion. (Rigorous proofs of these statements may be found in many references, e.g. Kushner [1974], Norman [1972], Watterson [1962].) These observations enable us to find the limiting diffusions of the processes arising in Genetics.

We list below several facts about diffusion processes that will be extensively used in the next section.

Let $\tau = \tau_{[a,b]}(x)$ be the first exit time from $[a,b]$ if the diffusion starts at x $(a < x < b)$. Then $T(x) = E\tau_{[a,b]}(x)$ solves

$$LT(x) \equiv \frac{1}{2}\sigma^2(x)\frac{d^2T(x)}{dx^2} + \mu(x)\frac{dT(x)}{dx} = -1,\ T(a) = T(b) = 0$$

(compare Gihman and Skorohord [1972], see also the appendix). As this differential equation can be reduced to a first order linear one its solution is easily found to be

$$T(x) = -\frac{2(S(b) - S(x))}{S(b)}\int_a^x \frac{S(\xi)d\xi}{\sigma^2(\xi)s(\xi)} + \frac{2S(x)}{S(b)}\int_x^b \frac{(S(b)-S(\xi))d\xi}{\sigma^2(\xi)s(\xi)}$$

where

$$s(x) = \exp\left\{-2\int_a^x \frac{\mu(\xi)}{\sigma^2(\xi)}d\xi\right\},\quad S(x) = \int_a^x s(\xi)d\xi\ .$$

Observe that if $\sigma^2(x) > 0$ over [a,b], then $T(x)$ is finite there, hence, the exit time is finite. Such an interval is called regular.

One can also find the Green function, $G(x,y)$, for the diffusion, where $G(x,y)$ is the expected time (per time unit) the process spends at y. Hence

$$E_x \int_0^\tau h(x_t)dt = \int G(x,y)h(y)dy \quad .$$

(In particular to get $E\tau$ take $h(x) \equiv 1$.) $G(x,y)$ solves $L_x G(x,y) = -\delta(x-y)$ and using standard methods in differential equations (compare Coddington and Levinson [1955], Birkhoff and Rota [1959]) one can show that

$$G(x,\xi) = \begin{cases} \dfrac{2S(x)\ (S(b) - S(\xi))}{S(b)\sigma^2(\xi)s(\xi)}, & \xi > x, \\[2ex] \dfrac{2(S(b) - S(x))S(\xi)}{S(b)\sigma^2(\xi)s(\xi)}, & \xi < x. \end{cases}$$

The hitting distribution on the boundary of such an interval, namely

$$u(x) = P_x\{X(\tau) = b\} = P_x\{\text{Reaching b before a}\},$$

satisfies $Lu = 0$, $u(a) = 0$, $u(b) = 1$. Hence $u(x)=S(x)/S(b)$.

A boundary point is called singular if $\sigma^2(x)$ vanishes there. If [a,b] is an interval where $\sigma^2(x) > 0$ for any x in its interior, but $\sigma^2(x)$ vanishes at one (or both) of the boundaries, several phenomena could happen. The process might reach the boundaries in finite time and become absorbed there, such boundaries are called exit, or else the process never reaches the boundary. The term used for such a boundary is unattainable (or inaccessible). Unattainable boundaries are further classified as attracting or repelling according to whether the process might converge to such a point with positive probability.

The following are very simple necessary and sufficient conditions for determining the nature of a boundary. First, note that b is repelling if for every $a < x < b_j \to b$

$$\Pr\{X(t) \text{ reaches } b_j \text{ before } a \,|\, X(0) = x\} \to 0 \quad (b_j \to b),$$

namely, if $u(x) \equiv 0$, equivalently, if $S(x)$ diverges near $x = b$ (i.e. $s(x)$ is not integrable there). Note that attractability is essentially stochastic stability. To determine if a boundary is attainable it is enough to show that the expected hitting time of that boundary is finite. (This is so because the distribution of the hitting time is easily seen to be bounded geometrically.) The expected hitting time of b, if the process starts at x and a acts as reflecting boundary, is given by

$$T^*(x) = 2\int_x^b s(u) \int_a^b \frac{dz}{\sigma^2(z)s(z)} \ .$$

So finiteness of the last expression is a necessary and sufficient condition for attainability of b.

If the two boundaries are repelling, then the process is recurrent, hence, it might have a stationary distribution. If a stationary distribution exists its density should satisfy the forward Kolmogorov equation when $\partial P/\partial t$ is set equal to zero. Maruyama and Tanaka [1957] proved indeed that this is the case for the one-dimensional homogeneous diffusions. Namely, a stationary distribution exists if and only if

$$p(x) = [\sigma^2(x)s(x)]^{-1} \in L^1[a,b] ,$$

and in that case $p(x)$ is the density of the stationary distribution (up to a multiplicative constant).

The theory of singular diffusions is due to Feller [1954] who used semi-groups arguments. A simple approach via differential equations to this theory is given in Prohorov and Rozanov [1969].

3. Applications to Genetics. The Wright-Fisher Models.

A) Constant selection coefficients.

Consider a finite population, say of size N, of two haplotypes A and B. Each individual gives birth to a large number of offsprings. From the large pool of young offsprings nature samples randomly N individuals who form the next generation. Thus in the Wright-Fisher model Z_n, the number of A alleles in the nth generation, evolves according to the following Markov chain

$$P_{ij} = P(Z_{n+1} = j | Z_n = i) = \binom{N}{j} p_i^j (1 - p_i)^{N-j} ,$$

where the p_i's depend on the relative fitness. More specifically if on the average an A individual gives birth to N_A individuals (and a B type to N_B individuals), then applying a deterministic approach to the large pool of young offsprings the proportion of A's will be $p_i = (1 + S)i/(N + Si)$, where $1 + S = N_A/N_B$.

There are standard techniques in Markov chain theory for getting fixation probabilities and expected times to absorption by solving appropriate systems of linear equations. However for most cases no explicit closed formulas exist. Hence one has to solve numerically for each set of parameters. Diffusion approximations provide us with approximate answers in terms of simple closed formulas.

To get a diffusion approximation for the W-F model let the time between transitions be $\frac{1}{N}$ $(h = \frac{1}{N})$ (namely N generations form one time unit) and $X_n^N = Z_n/N$. Then, in the no selection case,

$$\mu(x) = \lim_{h \to 0} \frac{1}{h} E\{X_{n+1}^N - X_n^N | X_n^N = x\} = 0 \quad (\text{as } E(X_{n+1}|X_n) = X_n) ,$$

$$\sigma^2(x) = \lim_{h \to 0} \frac{1}{h} \operatorname{Var}\{\Delta X_n | X_n^n = x\} = \lim_{N \to \infty} N \operatorname{Var}\{X_{n+1} | X_n = x\}$$

$$= \lim_{N \to \infty} N \frac{1}{N} x(1 - x) = x(1 - x).$$

Hence $X_n^N(nh) \to X(t)$ as $nh = n\frac{1}{N} \to t$, where $X(t)$ is the diffusion process with $\mu(x) = 0$, $\sigma^2(x) = x(1 - x)$. To arrive at a diffusion limit for the case where the relative fitness of A is $(1 + S) : 1$, we have to impose

$$S = S_h = sh = sN^{-1} \quad (s \text{ constant}),$$

thus, obtain $\mu(x) = sx(1 - x)$, $\sigma^2(x) = x(1 - x)$.

Using these limiting diffusion and the results of the first section, we easily see that in the no-selection case $u(x) = x$, $T(x) = -2x \ln x - 2(1 - x)\ln(1 - x)$ and $G(x,\xi) = 2x/\xi$ for $\xi > x$, while for $\xi < x$, $G(x,\xi) = 2(1 - x)/(1 - \xi)$, and in the selective case $u(x) = (\exp(-2sx) - 1)/(\exp(-2s) - 1)$. Note that $T(x)$ is given in terms of diffusion time units (one such unit corresponds to N generations). So the expected absorption time is $NT(x)$. The expected number of visits to j/N is approximated by $G(x, j/N)$ generations. Indeed

$$N \int G(x,\xi) \, \chi_{[j/N - 1/2N, \; j/N + 1/2N]}(\xi) d\xi \sim G(x,j/N)$$

generations ($\chi_A(z) = 1$ if $z \in A$, $\chi_A(z) = 0$ if $z \notin A$).

Let us examine how well these quantities approximate the exact ones. If there is no selection, Z_n/N is clearly a martingale (on $0, 1/N, 2/N, \cdots, (N-1)/N, 1$), hence, by the optional sampling theorem, $EZ_n = EZ_0 = 1/N = x = EZ_\tau = \Pr\{ Z_\tau /N = 1\}$. Namely, the fixation probability is x, thus, the diffusion approximation for this probability is exact. Consider now the selective case. One may prove that the diffusion approximation

$$u_s(x) = (\exp(-2\,sx) - 1)/(\exp(-2s) - 1)$$

is an upper bound for the exact fixation probability. Indeed let Z_n^* have the W-F transition matrix with $p_i^* = (1 - \exp(-2Si/N))/(1 - \exp(-2S))$. Using the fact that $\psi_n = \exp(-2SX_n)$ is a martingale, one easily verifies that $\Pr^*(A \text{ fixation} | X_0 = i/N) = (\exp(-2Si) - 1)/(\exp(-2S) - 1)$. However, as $p_i^* \geqslant p_i = (1 + S)i/(N + Si)$, $S > 0$, Z_n^* is stochastically larger than Z_n and, therefore, the probability that X_r^* reaches 1 is an upper bound to the fixation probability of A. (These observations are due to Moran [1960].)

The diffusion approximation for the fixation probability, though an upper bound for the exact quantity, is an extremely good approximation for small s. Let us indicate how to improve this approximation (for moderate s). Expanding

$\exp(-2s\ X_{t+1})$ around X_t and using the fact that $\Delta X_t = X_{t+1} - X_t$ is Binomially distributed, one obtains

$$E\{\exp(-2s\ X_{t+1})\} = E\{\exp(-2s\ X_t) + (1/N^2)h(X_t)\} + O(1/N^2),$$

where $h(x) = (2/3N^2)s^3x(1 - x^2)\exp(-2s\ x)$. Integrating this, one obtains

$$E_x\{\exp(-2s\ X_{t+1})\} = \exp(-2sx) + E_x\{(1/N^2)\sum_{j=0}^{t} h(X_j) + tO(1/N^2)\}.$$

Using the optional sampling theorem we can replace $t + 1$ by τ, the absorption time, thus, obtain

$$\begin{aligned} E_x\exp(-2sX_\tau) &= \exp(-2sx) + (1/N^2)E\sum_{j=0}^{\tau-1} h(X_j) + E(\tau)O(1/N^2) \\ &= \exp(-2sx) + (1/N^2)\sum_{k=1}^{N-1} h(\tfrac{k}{N})G(\tfrac{i}{N}, \tfrac{k}{N}) + E(\tau)O(1/N^2), \quad x = i/N, \end{aligned}$$

where $G(\frac{i}{N}, \frac{k}{N})$ is the Green function for X_n; namely, $G(\frac{i}{n}, \frac{j}{n})$ is the expected number of visits to j/N if $x_0 = i/N$, thus, $G(i/N,j/N) = \sum_{\ell=0}^{\infty} P^{\ell}_{ij}$.

Replacing the Green function of the original process by its diffusion approximation (hence, approximating $(1/N)\sum_{k=1}^{N-1} h(\frac{k}{N})G(\frac{i}{N}, \frac{k}{N})$ by an integral), we obtain the following improvement for the diffusion approximation of the fixation probability

$$u(x) = \frac{\exp(-2sx) - 1}{\exp(-2s) - 1} - \frac{1}{N}\frac{A + B}{1 - \exp(-2s)},$$

where

$$A = \frac{1}{6}\,\frac{\exp(-2s)-\exp(-2sx)}{\exp(-2s)-1}\,[4s^2(x + \tfrac{1}{2}x^2) + 2s\{(1+x)\exp(-2sx)-1\}],$$

$$B = \frac{1}{6}\,\frac{\exp(-2sx)-1}{\exp(-2s)-1}\,[\exp(-2sx)\{1 + 2sx + 2s\}$$

$$- \exp(-2s)\{6s^2 - 4s^2x - 2s^2x^2 + 4s + 1\}].$$

It turns out that this improved approximation due to Ewens [1964] is extremely good.

It should be noted that a similar method for improving the expected time formula fails. The reason being that $T(x) = -2x \ln x - 2(1-x)\ln(1-x)$ has unbounded derivatives at $x = 0$ and $x = 1$. However comparing the diffusion approximation with numerical results for the Green function (obtained by solving a system of linear equations), one concludes that the diffusion approximation for $G(i/N, j/N)$ is quite good; it usually overestimates the exact value by 3-5 percent. However if $i = j$, then the diffusion approximation for $G(i/N, i/N)$ is 2, while the exact value lies between 2.7 to 3 in most cases. Moreover the diffusion approximation (for $G(i/N, j/N)$) is not too good if j/N is close to either boundary. (Compare e.g. Ewens [1964], Pollak and Arnold [1975].)

B) Varying Deterministic Selection.

Consider now the case where selection varies deterministically with time, say $S = S(n) = Sf_N(n)$. Thus Z_n is governed by the following non-homogeneous Markov chain,

$$P_{ij}(n) = \Pr(Z_{n+1} = j | Z_n = i) = \binom{N}{j} p_i^j(n)(1 - p_i(n))^{N-j},$$

where $p_i(n) = i(1 + Sf(n))/(N + iSf(n))$. To find $u_i(n)$, the probability of ultimate fixation if at time n A has frequency i, one has to solve the following system of linear equations

$$u_i(n) = \sum_{j=0}^{N} P_{i,j}(n) u_j(n+1), \quad u_0(n) = 0, \quad u_N(n) = 1, \quad n \geq 0,$$

$$i = 0, 1, \ldots N,$$

which is quite difficult.

Thus we resort to diffusion approximations. It is easily established that if $f_N(n) \to f(t)$ as $n/N \to t$, then X_n^h converges to a non-homogeneous diffusion with $\mu(x,t) = sf(t)x(1 - x)$ and $\sigma^2(x) = x(1 - x)$. $u(x,t)$, the probability of ultimate fixation if $X(t) = x$, can be shown to satisfy (see the appendix)

$$-\frac{\partial u}{\partial t} = \frac{1}{2}\sigma^2 \frac{\partial^2 u}{\partial x^2} + \mu \frac{\partial u}{\partial x} = L_x u; \; u(0,t) = 0, \; u(1,t) = 1 \quad (t \geq 0).$$

It has been proved (Levikson and Schuss [1976]) that this equation has a unique solution. Moreover if $f(t) \to \gamma$ $(t \to \infty)$, then $u(x,t) \to u(x)$ $(t \to \infty)$, where $u(x)$ is the solution of the ordinary differential equation

$$\frac{1}{2}x(1 - x)u_{xx} + \gamma x(1 - x)u_x = 0, \quad u(0) = 0, \quad u(1) = 1.$$

To solve the above partial differential equation numerically let $u_n(x,T_n) = v(x)$, where $v(x)$ is an arbitrary function and T_n an increasing sequence, and employ the forward scheme:

$$u_n(x,t-\Delta t) = u_n(x,t) + \Delta x\ s\ f(t)x(1-x)(u_n(x+\Delta x,t)-u_n(x,t)) + (1/2)x(1-x)(u_n(x+\Delta x,t)-2u_n(x,t)+u_n(x-\Delta x,t));$$

$$t = T_n, \quad T_n = -2\Delta t,\ldots,2\Delta t, \quad x = \Delta x, 2\Delta x, \ldots, 1 - \Delta x, 1,$$

where $\Delta t = (\Delta x)^2$, to obtain $u_n(x,0)$ eventually. Once $u_n(x,0)$ stabilizes it is taken to be the fixation probability. Several examples, where $f(t) = e^{-t}$ and where $f(t) = (1+t)^{-1}$, $s = 0.5, 1.0, 1.5$, $\Delta x = 0.05$, $\Delta t = 0.0025$, show that the scheme is extremely stable; namely, when we put $u_1(x,10) = x$ and $u_2(x,20) = x$ as initial conditions, $u_1(x,0)$ and $u_2(x,0)$ turn out to be the same for the first 4 significant digits.

Another way to find $u(x,t)$ is to solve the partial differential equation by power series. Thus if, for example, $f(t) = e^{-t}$ letting $u(x,t) = \sum_{i=0}^{\infty} u_i(x)e^{-it}$, one obtains a recursive system of ordinary differential equations for the u_i's which are readily solved. It turns out that $u_i(x)$ has the form $u_i(x) = s^i\overline{u}_i(x)$, where $\overline{u}_i(x)$ is independent of

s, so that for small values of s ($|s| < 1$) only the first few terms of the power series are significant. In particular $u_0(x) + u_1(x)e^{-t}$ turns out to be extremely close to the numerical result for $u(x,t)$ (compare Levikson and Schuss [1976]).

Next let us consider random selection. Assume the selection intensities form a sequence of independent random variables. Then the transition matrices themselves are random and, consequently, it is very difficult to find the fixation probabilities for this case, hence, once again we apply diffusion approximations. Let the selection scheme in the nth generation be $1 + S_1(n) : 1 + S_2(n)$ and let $ES_i(n) = s_i/N$, $ES_i^2(n) = v_i/N$, $ES_1(n)S_2(n) = r/N$ and all the other moments be $o(1/N)$. Then X_n^h converges weakly to the diffusion with $\mu(x) = x(1-x)(s_1-s_2+v_1x+v_2(1-x)+r(2x-1))$ and $\sigma^2(x) = x(1-x)[1 + (v_1+v_2-2r)x(1-x)]$ (see Levikson [1974]). Using this diffusion approximation we are able to deduce that the fixation probability of an "advantageous" allele decreases as the variance of its selective forces increases (keeping the mean selection constant). Here by the "advantageous" allele we mean the one with the highest initial frequency if the two selective forces have the same distribution, while in the asymmetric case, where the relative selective force of A with respect to B is $1 + S:1$, A is considered "advantageous" if $ES > 0$. We have also shown that in the symmetric case the expected time to absorption decreases as the variance increases, while for the asymmetric case this is true provided the initial A frequency is not too close to 1. The proofs of these statements and many additional results on this case may be found in Levikson [1974]. (See also Karlin & Levikson [1974] for the statements of the main results.) Earlier works are due to Jensen and Pollak [1969] and Jensen [1973].

C) Wright Fisher Models with Mutation: Two types of diffusion approximations.

Assume that A mutates into B with probability α_1 and B mutates into A with probability α_2. Then the W-F chain is clearly regular. Hence, standard Markov chain theory implies that the process attains a stationary distribution π satisfying $\pi P = \pi$, $\pi \geqslant 0$, $(\pi, e) = 1$, $e = (1,\ldots,1)$. No explicit formulas are known for π. To get an idea on the behavior of π, observe that $EX_{t+1} = (1 - \alpha_1)EX_t + \alpha_2(1 - EX_t)$. Hence the expectation of the stationary distribution, m, should satisfy $m = (1 - \alpha_1)m + \alpha_2(1 - m)$; namely $m = \alpha_2(\alpha_1 + \alpha_2)^{-1}$. Similar computations for the variance of the stationary distribution yield that it is of the order N^{-1} (compare Moran [1962]), i.e. for large N the stationary distribution is quite concentrated around its mean.

To get an approximation for the stationary distribution via diffusions one therefore has to impose $N\alpha_i \to \beta_i$, thus, obtaining a limiting diffusion with $u(x) = -\alpha_1 x + \alpha_2(1 - x)$, $\sigma^2(x) = x(1 - x)$. Using the observations made in the first section we conclude that this diffusion converges to a stationary Beta-type distribution whose density is given by

$$f(x) = \frac{\Gamma(2(\beta_1+\beta_2))}{\Gamma(2\beta_1)\Gamma(2\beta_2)} x^{2\beta_2-1}(1 - x)^{2\beta_1-1} \qquad (0 < x < 1).$$

Note that if both mutation rates are large (to the extent that $2\beta_i < 1$), this density has an "open down U" shape indicating that we are likely to find a fair amount of individuals of each type. However if $2\beta_i > 1$ $(i = 1,2,)$, the stationary density has an "open upward U" shape, meaning we are likely to find either many A's and very few B's or vice-versa.

As was earlier indicated, the A gene frequency tends in probability to $m = \alpha_2/(\alpha_1 + \alpha_2)$ as the population increases, indeed, increasing N means that the β's of the diffusion limit become large and so the stationary distribution gets more and more concentrated around the "quasi" fixed point m.

To study the fluctuations around the quasiequilibrium m when the mutation rates are quite large, a Gaussian type diffusion approximation is most appropriate. This approximation is obtained if we let $\alpha_i = \beta_i h$, $Nh \to \infty$ and consider $(X_n^h - m)\sqrt{N}$, whose limit is a diffusion on the real line with $\mu(x) = -(\beta_1 + \beta_2)x$, $\sigma^2(x) = k$, $(-\infty < x < \infty)$. This process is known as the Orenstein-Uhlenbeck process and it can be characterized as the only stationary Gaussian diffusion. It was first introduced to Genetics by Feller [1951]. Later Karlin & McGregor [1964] obtained Gaussian approximations to Moran's models and Norman [1975] rigorously derived these approximations. Moreover Norman [1975 b] proved the convergence of the transition density to the appropriate stationary density directly, without using diffusion approximations!

4. RANDOM SELECTION IN LARGE POPULATIONS: DIFFUSION APPROXIMATIONS WITH UNATTAINABLE BOUNDARIES.

Consider an infinite diploid population having the genotypes AA, AB, BB, whose selection coefficients in the nth generation are $1 + S_1(n): 1: 1 + S_2(n)$, respectively. Then p_n, the proportion of the A gene in the nth generation, follows the rule

$$p_{n+1} = \frac{p_n^2(1 + S_1(n)) + p_n(1 - p_n)}{1 + p_n^2 S_1(n) + (1 - p_n)^2 S_2(n)}.$$

Thus we are facing the problem of finding the limiting behavior of stochastic iterations. One way to deal with random iterations is linearizing the recursive equation around its fixed points, studying the linearized process, and inferring from it to the original one. (See Bailey [1964], Gillespie [1973], Karlin & Liberman [1974] and Levikson [1976] for details and examples on this method.) Another way is using diffusion approximations.

To obtain a diffusion approximation let the time between transitions be h and let $ES_i = s_i h$, $ES_i^2 = v_i h$, $ES_1 S_2 = rh$ and all the other moments be $o(h)$. Then the gene frequency process converges to a diffusion with

$$\mu(x) = x(1-x)[s_1 x - s_2(1-x) - v_1 x^3 + v_2(1-x)^3 + rx(1-x)(2x-1)],$$

$$\sigma^2(x) = x^2(1-x)^2[v_1 x^2 + v_2(1-x)^2 - 2rx(1-x)].$$

To study the limiting diffusion, observe that $x = 0$ is attracting if and only if $2s_2 > v_2$. Indeed, since $s(x) \sim Kx^{2s_2/v_2-2}$ near $x = 0$, $2s_2 > v_2$ is a necessary and sufficient condition for $s(x)$ to be integrable there. (Similarly, $x = 1$ is attracting if and only if $2s_1 > v_1$.) Moreover a stationary distribution exists if and only if $2s_1 < v_1$ and $2s_2 < v_2$. Indeed, $f(x) = [\sigma^2(x)s(x)]^{-1}$ is continuous in $(0,1)$ and, since $f(x) \sim Kx^{-2s_2/v_2}$ near $x = 0$, it is integrable there provided $2s_2 < v_2$. (Similarly $f(x)$ is integrable near $x = 1$ if and only if $2s_1 < v_1$.)

Thus we have the following limiting possibilities.

(i) If $2s_2 > v_2$, $2s_1 \leqslant v_1$, then $x = 0$ is attracting, $x = 1$ is repelling and the A gene frequency converges to $x = 0$.

(ii) If $2s_2 \leqslant v_2$, $2s_1 > v_1$, then $x = 1$ is attracting, $x = 0$ is repelling and the A gene is going to be fixed.

(iii) If $2s_2 > v_2$, $2s_1 > v_1$, then both boundaries are attracting, hence, in a certain proportion of the sample paths the A gene is going to be fixed while in the remaining ones it is going to be extinct.

(iv) If $2s_1 < v_1$, $2s_2 < v_2$, then both boundaries are repelling and the process attains a stationary distribution.

If $2s_2 = v_2$, then $x = 0$ is repelling, but a stationary distribution fails to exist. In such a case, the diffusion is recurrent, but the mass of the distribution piles up at the origin.

It should be noted that under our moments conditions on the S_i's the requirement $2s > v$ is equivalent to $E \log(1 + S) > 0$. Indeed, the conditions for local stochastic stability in the discrete case are given in terms of $E \log(1 + S)$.

(The above results are due to Levikson [1974]; see also Levikson and Karlin [1975]. Earlier studies on this problem via diffusions are due to Gillespie [1973 a], Hartl and Cook [1973], Kimura [1954].)

APPENDIX. First let us derive the partial differential equation for the fixation probability. Let $u_{[a,b]}(x,t)$ be the probability that $X(s)$ $(s \geq t)$ reaches b before a if at time t the process is at x $(a \leq x \leq b)$. The Markov property implies

$$u(x,t) = \int p(x,t,y,t+h)u(y,t+h)dy.$$

Expanding $u(y,t+h)$ around (x,t) and using the basic assumptions on the infinitesimal moments of diffusion processes, we obtain

$$u(x,t) = \int p(x,t,y,t+h)(u(x,t)+u_x(x,t)(y-x) + u_t(x,t)h$$

$$+ \tfrac{1}{2}u_{xx}(x,t)(y-x)^2 + o(y-x)^2 + o(h))dy$$

$$= u(x,t) + u_x(x,t)\mu(x,t)h + u_t(x,t)h$$

$$+ u_{xx}(x,t)\sigma^2(x,t)h + o(h).$$

Hence $u(x,t)$ satisfies

$$-\frac{\partial u}{\partial t} = Lu \equiv \frac{1}{2}\sigma^2(x,t)\frac{\partial^2 u}{\partial x^2} + \mu(x,t)\frac{\partial u}{\partial x},$$

$$u(a,t) = 0, \quad u(b,t) = 1 \quad (t \geqslant 0).$$

For a rigorous proof using Ito's rule we refer the reader to Gihman and Skorohod [1972].

Similarly $T(x,t)$, the expected time to absorption if $X(t) = x$, satisfies

$$-\frac{\partial T}{\partial t} - 1 = LT, \quad T(a,t) = T(b,t) = 0.$$

Indeed observe that

$$T(x,t) = h + \int p(x,t,y,t+h)T(y,t+h)dy$$

and proceed as before.

References

1. N. T. J. Bailey, *The Elements of Stochastic Processes,* J. Wiley, New York (1964).

2. A. T. Bharucha-Reid, *Elements of the theory of Markov Processes and their Applications*, McGraw Hill, New York (1960).

3. G. Birkhoff and G.-C. Rota, *Ordinary Differential Equations*, Ginn & Co., Boston (1959).

4. E. A. Coddington and N. Levinson, *Theory of Ordinary Differential Equations*, McGraw Hill, New York (1955).

5. W. J. Ewens, *The Pseudo-transient Distribution and its Uses in Genetics*, J. Appl, Prob., 1 (1964), 141-156.

6. W. Feller, *Diffusion Processes in Genetics*, 2nd Berkeley Symp. Math. Stat. and Prob. (1951), (J. Neyman Ed.) Univ. of Cal. Press, 227-246.

7. W. Feller, *Diffusion Processes in One Dimension,* Trans. Am. Math. Soc., 77 (1954), 1-31.

8. I. I. Gihman and A. V. Skorohod, *Stochastic Differential Equations,* Springer Verlag, Berlin (1972).

9. J. Gillespie, *Polymorphism in Random Environments*, Theor. Pop. Biol., 4 (1973), 193-195.

10. J. Gillespie, *Natural Selection with Varying Selection Coefficients - a Haploid Model*, Genet. Res. 21 (1973 a), 115-122.

11. H. A. Guess, *On the Weak Convergence of Wright-Fisher Models*, J. of Stochastic Processes, 1 (1973), 287-306.

12. D. Hartl and D. Cook, *Balanced Polymorphism of Quasi-Neutral Alleles*, Theor. Pop. Biol, 4 (1973), 163-172.

13. L. Jensen, *Random Selective Advantages of Genes and Their Probability of Fixation*, Gen. Res., 21 (1973) 215-219.

14. L. Jensen and E. Pollak, *Random Selection Advantages of a Gene in Finite Populations*, J. App. Prob. 6 (1969), 19-37.

15. S. Karlin and B. Levikson, *Random Selection Intensities: Case of Small Populations*, Theor. Pop. Biol 6 (1974), 383-412.

16. S. Karlin and U. Liberman, *Random Selection Intensities: Case of Large Populations*, Theor. Pop. Biol 6 (1974), 355-382.

17. S. Karlin and J. McGregor, *On Some Stochastic Models in Genetics*, Stochastic Models in Medicine and Biology (J. Gurland Ed.), Univ. of Wisc. Press, Madison (1964), 245-271.

18. M. Kimura, *Process Leading to Quasi-Fixation of Genes*

in Natural Population due to Random Fluctuations of Selection Intensities, Genetics 39 (1954), 280-295.

19. H. T. Kushner, *On the Weak Convergence of Interpolated Markov Chains to Diffusion*, Anal. Prob. 2 (1974), 40-50.

20. B. Levikson, Ph.D. Thesis, Tel Aviv Univ. (1974).

21. B. Levikson, *Regulated Random Growth*, J. Math. Biol., to appear.

22. B. Levikson and S. Karlin, *Random Selection Intensities in Large Diploid Populations*, Theor. Pop. Biol 8 (1975), 292-303.

23. B. Levikson and Z. Schuss, *Nonhomogeneous Diffusion Approximation to a Genetic Model*, J. Math. Pures. Appl., to appear.

24. G. Maruyama and H. Tanaka, *Some Properties of one Dimensional Diffusion Processes*, Men. Fac. Sci., Kyushu Univ. Series A 11 (1957), 117-141.

25. P. A. P. Moran, *The Survival of a Mutant Gene Under Selection II*, J. of Australian Math. Soc. 1 (1960), 485-491.

26. P. A. P. Moran, *The Statistical Process of Evolutionary Theory*, Clarendon Press, Oxford (1962).

27. M. F. Norman, *Markov Processes and Learning Models*, Academic Press, New York (1972).

28. M. F. Norman, *Approximation of Stochastic Processes by Gaussian Diffusions and Applications to Wright-Fisher*

Genetic Models, SIAM J. Appl. Math 29 (1975 a), 225-242.

29. M. F. Norman, *Limit Theorems for Stationary Distributions*, Advances in Appl. Prob. 7 (1975 b), 561-575.

30. E. Pollak and B. C. Arnold, *On Sojourn Times at Particular Gene Frequencies*, Genet. Res. (Camb.) 25 (1975) 89-94.

31. Yu. V. Prohorov and Ya. A. Rozanov, *Probability Theory*, Springer Verlag, New York (1969).

32. G. A. Watterson, *Some Theoretical Aspects of Diffusion Theory in Population Genetics*, Ann. Math. Stat. 33 (1962), 939-957.

Division of Mathematical Sciences
Purdue University
West Lafayette, Indiana

Some Unsolved Problems and New Problem Areas in the Field of Stochastic Differential Equations

BY

James McKenna

Abstract. This paper is of a tutorial nature. It has several purposes. Its most direct aim is to suggest some unsolved or partly solved problems in stochastic differential equations which might be of interest. A second aim is to point out to the reader the areas of technology from which these problems arise, for there are many more problems where these came from.

Specifically, a problem involving coupled mode equations is discussed in some detail. A class of problems is pointed out where it is desired to infer properties of the coefficients of the equations from knowledge of the solutions. Then a system of equations is shown where non-stochastic techniques have been employed to study the dependence of the solution on a parameter. Finally, brief mention is made of stochastic difference equations.

1. Introduction. I was asked by the organizers of the Conference on Stochastic Differential Equations to talk about

some interesting unsolved problems and/or possible new areas of research. Much of the motivation for studying stochastic differential equations came from the study of Brownian motion in physics. This has merged with the theory of Markov processes and forms a coherent and fairly well defined area of mathematics. There is however, another class of stochastic differential equations which arise in modern technology. Aside from their origin, they do not form, as yet, a well defined area of mathematics. However, many of them have immense practical importance, and I want to describe some of them here. Consequently, what follows is a description of unsolved problems, problem areas illustrated by specific solved problems, and brief comments about completely unsolved problems. All the problems I shall discuss arise from communications technology, which reflects my Bell Telephone Laboratories background. None of the work described is my own.

I will discuss three specific problems in some detail, and finally say a little about what might be an interesting new area of research. First, I will discuss a waveguide problem involving coupled mode equations. It is a problem of the direct type: the coefficient stochastic process is known and various moments of the solution are desired. Since information about this problem is not too widespread, I will be most expansive here. Second, I will talk about inverse problems: given some knowledge of the solution stochastic process, what information can you get about the coefficient stochastic process. I will discuss several areas of technology where such problems arise and then describe some interesting new results in this field due to B. Gopinath [1]. Third, I will discuss a problem where the dependence on a nonstochastic parameter of the solution of a system of stochastic differential equations is desired. The problem has been attacked in an elegant fashion by D. Mitra and M. M. Sondhi [2] using completely nonstochastic techniques. It seems reasonable to ask if there exist stochastic techniques which would yield the same results. Finally, I will mention some interesting stochastic difference equations arising from some new electronics technology.

2. COUPLED MODE EQUATIONS WITH RANDOM COUPLING.

The first problem I want to discuss arose in a study of mode coupling due to bends in the WT4 waveguide system by D.J. Thomson and D.T. Young of the Bell Laboratories. Interesting descriptions of the WT4 waveguide system can be found in [3,4]. This problem is a nice example of the many which come out of coupled mode theory. Since coupled mode theory [5,6] is an important tool of physical scientists and engineers which is perhaps not too familiar to mathematicians, I will spend some time developing the equations.

For our purposes here, such a waveguide can be assumed to be a long, hollow metal tube which can guide electromagnetic signals. To get to the mathematics, consider an infinitely long, hollow tube, constructed from a perfectly conducting metal, and having a uniform cross section, as shown in Figure 1. It is well known that there exist solutions of Maxwell's equations which represent waves contained in the interior of the tube, and which propagate along the axis of the tube with no loss of energy [7]. The analysis begins with Maxwell's equations which for our purposes here I can write as

$$\underset{\sim}{\nabla}\times\underset{\sim}{E} = -\mu_0 \frac{\partial \underset{\sim}{H}}{\partial t}, \quad \underset{\sim}{\nabla}\times\underset{\sim}{H} = \varepsilon_0 \frac{\partial \underset{\sim}{E}}{\partial t}. \tag{2.1}$$

In these equations $\underset{\sim}{E}$ and $\underset{\sim}{H}$ are the electric and magnetic field intensities, μ_0 and ε_0 are constants (the permeability and permittivity of the vacuum), and $\underset{\sim}{\nabla}$ is the gradient operator. We seek solutions of (2.1) of the form

$$\begin{aligned} \underset{\sim}{E}(x,y,z,t) &= \underset{\sim}{\mathcal{E}}(x,y)e^{i(\omega t-\beta z)}, \\ \underset{\sim}{H}(x,y,z,t) &= \underset{\sim}{\mathcal{H}}(x,y)e^{i(\omega t-\beta z)}, \end{aligned} \tag{2.2}$$

which on the sides of the guide satisfy the boundary conditions

$$\underset{\sim}{n}\times\underset{\sim}{\mathcal{E}} = 0, \qquad \underset{\sim}{n}\cdot\underset{\sim}{\mathcal{H}} = 0, \tag{2.3}$$

In (2.2), $i = \sqrt{-1}$, ω is the radian frequency and is given, β is the propagation constant and must be determined, and in (2.3), $\underset{\sim}{n}$ is the unit vector normal to the sides of the guide. When the expressions (2.2) are substituted into Maxwell's equations (2.1), there result the equations

$$\begin{aligned} \underset{\sim}{\nabla}_t \times \underset{\sim}{E} - i\beta \underset{\sim}{k} \times \underset{\sim}{E} &= - i\omega\mu_0 \underset{\sim}{H}, \\ \underset{\sim}{\nabla}_t \times \underset{\sim}{H} - i\beta \underset{\sim}{k} \times \underset{\sim}{H} &= i\omega\varepsilon_0 \underset{\sim}{E}. \end{aligned} \tag{2.4}$$

In (2.4)

$$\underset{\sim}{\nabla}_t = \underset{\sim}{i} \frac{\partial}{\partial x} + \underset{\sim}{j} \frac{\partial}{\partial y} \tag{2.5}$$

and $\underset{\sim}{i}$, $\underset{\sim}{j}$, $\underset{\sim}{k}$ are unit vectors in the x, y and z directions, respectively.

It is well known that the equations (2.4) with the boundary conditions (2.3) have a countable infinity of solutions, which I denote by $\underset{\sim}{E}_{(n)}$, $\underset{\sim}{H}_{(n)}$, $\beta_{(n)}$. Since Maxwell's equations are linear, the most general solution describing a wave in the guide moving in the positive z-direction has the form

$$\begin{aligned} \underset{\sim}{E}(x,y,z,t) &= \sum_{(n)} a_{(n)}(z) \underset{\sim}{E}_{(n)}(x,y) e^{i\omega t} \\ \underset{\sim}{H}(x,y,z,t) &= \sum_{(n)} a_{(n)}(z) \underset{\sim}{H}_{(n)}(x,y) e^{i\omega t}, \end{aligned} \tag{2.6}$$

$$a_{(n)}(z) = a_{(n)} e^{-i\beta_{(n)} z}, \tag{2.7}$$

where the $a_{(n)}$ are constants. The solution corresponding to each n is called a mode. Clearly, I can choose $a_{(m)} \equiv 1$,

$a_{(n)} \equiv 0$, $n \neq m$, which in engineering parlance says that a single mode can propagate uncoupled to any other modes.

The set of coefficient functions $a_{(n)}(z)$ is just a solution of the system of differential equations

$$\frac{da_{(n)}(z)}{dz} + i\beta_{(n)}a_{(n)}(z) = 0. \tag{2.8}$$

Now so far I have only discussed the waves which can propagate in a perfectly straight waveguide. However, in 1957 Morgan studied wave propagation in waveguides whose axes are bent [8]. He showed that the expression for the most general field in the guide is still given by (2.6) with the same functions $\underset{\sim}{E}_{(n)}$ and $\underset{\sim}{H}_{(n)}$. But now the $a_{(n)}(z)$ are the solution to a set of coupled differential equations of the form

$$\frac{da_{(n)}(z)}{dz} = \sum_{(m)} C_{(n),(m)}(z)a_{(m)}(z). \tag{2.9}$$

The quantities $C_{(n),(m)}(z)$ can be calculated explicitly and are functions of the curvature of the axis of the guide. Now even if we pick initial conditions so that at $z = 0$ there is just a single mode moving in the positive z-direction, say $a_{(m)}(0) = 1$, $a_{(n)}(0) = 0$, $n \neq m$, because of (2.9), $a_n(z) \neq 0$, $n \neq m$ in general for $z \neq 0$. In other words, the mth mode is now coupled to many other modes moving in both directions.

In general, it is impossible to solve the set of equations (2.9) exactly. However, in many cases it is possible to set to zero all but a finite number of the $C_{(n),(m)}(z)$ in (2.9) and the solutions of the resulting finite set of "coupled mode" equations accurately describes the system of interest.

I have now reached a specific problem of interest: consider the coupling of a desired mode $a_0(z)$ to a single undesired mode $a_1(z)$ due to bends in the waveguide. Write the two coupled mode equations in the form

$$\frac{da_0}{dz} + i\beta_0 a_0 = iC_{01}K(z)a_1,$$
$$\frac{da_1}{dz} + i\beta_1 a_1 = iC_{01}K(z)a_0. \tag{2.10}$$

The quantities β_n are the propagation constants of the modes in the absence of any coupling. The constants C_{01}, β_0 and β_1 come out of Morgan's theory [8] and $K(z)$ is the curvature of the axis of the guide. (The axis is assumed to be a plane curve.) It is impossible to fabricate a waveguide with a perfectly straight axis; the axis will have small, random bends in it. Thus we must assume that $K(z)$ is a stochastic process. What I want to know is the following. If the initial conditions

$$a_0(0) = 1, \qquad a_1(0) = 0 \tag{2.11}$$

are imposed, that is, if the desired mode is excited at $z = 0$, how far down the waveguide can I go before the undesired mode grows to objectionable size? Since this is a stochastic problem, I want to calculate $\langle a_0(z) \rangle$ and $\langle a_1(z) \rangle$, where $\langle\ \rangle$ denotes the stochastic average.

The problem as I have formulated it seems quite standard. This case is of special interest, however, because for once in a problem of this kind, a great deal is known about the stochastic coefficient $K(z)$. Very careful measurements of $K(z)$ have been made on a test installation of the WT4 by Thomson, Young and their colleagues, and they have 1.36×10^4 meters of sample function. They have analyzed their data very carefully and concluded that $K(z)$ is a stationary Gaussian processes with a correlation length of about 4.5 meters. They cut their 13.6 kilometers of sample function into 16 independent sample functions each 800 meters long. They then numerically calculated the solutions of (2.10) for each of these sample functions and then calculated $\langle a_0(z) \rangle$ and $\langle a_1(z) \rangle$. Next they calculated $\langle a_0(z) \rangle$ and $\langle a_1(z) \rangle$ using several types of perturbation theory [9,10]. The results of perturbation theory did not agree well with the numerical results in some interesting parameter ranges!

To get some hints about what is going sour, expand $a_0(z)$ and $a_1(z)$ in power series in C_{01}. The first few terms are

$$a_0(z) = e^{-i\beta_0 z}\left\{1 - (C_{01})^2 \int_0^z du \int_0^u e^{-i\Delta\beta(u-v)}K(u)K(v)dv + \dots\right\},$$

$$a_1(z) = e^{-i\beta_0 z}\left\{iC_{01} \int_0^z e^{-i\Delta\beta(z-u)}K(u)du + \dots\right\}, \tag{2.12}$$

where $\Delta\beta = \beta_1 - \beta_0$. If I set

$$\gamma(u-v) = \langle K(u)K(v)\rangle,$$

$$S(f) = \int_{-\infty}^{\infty}\gamma(u)e^{-2\pi iuf}du,$$

$$T(f) = \int_0^1(1-v)e^{-2\pi ifv}dv = \frac{1}{2i}\left\{\frac{\pi f - e^{-i\pi f}\sin \pi f}{\pi^2 f^2}\right\},$$

then it is easy to show that

$$\langle a_0(z)\rangle \approx e^{-i\beta_0 z}\left\{1 - (zC_{01})^2 \int_{-\infty}^{\infty}T(fz)S\left(\frac{\Delta\beta}{2\pi} - f\right)df\right\}. \tag{2.13}$$

Now $T(f) \sim \frac{1}{2\pi if}$ as $f \to \infty$, while $T(f) \sim i$ as $f \to 0$, so the main contribution of the integral in (2.13) occurs near $f = 0$. A rough graph of $S(f)$ is given in Figure 2, and it should be noted that $S(f)$ drops by four orders of magnitude in $0 \leqslant f \leqslant .6\ m^{-1}$. For the problem of interest, $\Delta\beta/2\pi \sim (1/3)m^{-1}$, $C_{01} \sim 10$ to 24 and $S(1/3) \sim 2\times10^{-7}m^{-1}$. At $f \sim .6\ m^{-1}$, where $S(f) \sim 10^{-8}$, standard perturbation theory works very well, but if you get just a little way up the shoulder, it starts to break down. What approximate technique will work well here?

3. INVERSE PROBLEMS. In the previous section I discussed the problem of the direct type: the stochastic process used as a coefficient was known and some properties of the solution stochastic process were desired. However, the problem of section 2 was somewhat unique because for once a great deal is known about the coefficient stochastic process. In fact, in many important cases the inverse problem is of more interest. Given a certain amount of information about the solution of a stochastic differential equation, what can you say about the coefficient stochastic process? In many of these cases, some knowledge of coefficients is also available, but more is needed. I think this is one of the most challenging areas in the field of stochastic differential equations. The problem is a difficult one, although some recent progress has been made. Here are a few examples.

Consider the following simple problem. An electronics manufacturer wishes to produce a linear R-C circuit [11] such that $\tau_0 = R_0C_0$ seconds after a unit step voltage is applied to the circuit, the voltage across the capacitor has decreased to $1-e$ (where $e = 2.7182$). The manufacturer can only hope to produce components whose values are, on the average, uniformly distributed in the intervals $(1-\varepsilon)R_0 \leqslant R \leqslant (1+\varepsilon)R_0$,

$(1-\delta)C_0 \leqslant C \leqslant (1+\delta)C_0$. How small must δ and ε be so that the manufacturer can guarantee that in some average sense for each circuit he produces, τ_0 seconds after a unit voltage step is applied the voltage across the capacitor will be in some interval about $1-e$?

This problem can be translated into mathematics as follows. The voltage $v(t)$ across the capacitor after the application of the unit step voltage is the solution of the differential equation

$$RC \frac{dv}{dt} + v = 1 \tag{3.1}$$

which satisfies the initial condition $v(0) = 1$. The quantities R and C are independent random variables uniformly distributed on the intervals $R_0(1-\varepsilon) \leqslant R \leqslant R_0(1+\varepsilon)$, $C_0(1-\delta) \leqslant C \leqslant C_0(1+\delta)$. We can, for example, calculate $m = \langle v(\tau_0)\rangle$ and $\sigma^2 = \langle\{v(\tau) - \langle v(\tau)\rangle\}^2\rangle$ as functions of δ and ε. We could then choose δ and ε so that σ and $|m-1+e|$ are suitably small. Alternatively, we could calculate $P(\delta,\varepsilon;u) = \text{Prob}\{|v(\tau_0) - 1+e| > u\}$. For given u, we could then pick ε and δ so that $P(\delta,\varepsilon;u)$ is suitably small.

This is a trivial example of a class of problems of great importance in electronics technology. In most cases, Monte Carlo techniques are used, but it seems to me that this area of statistical circuit design will provide many interesting problems for researchers in stochastic differential equations. The interested reader should consult [12] for a number of survey papers on this subject.

Another interesting inverse problem is provided in some recent work by B. Gopinath [1] which has applications to the study of thin optical films. Consider a stack of n layers of dielectric material, each layer of thickness ℓ_i and dielectric constant K_i, $i = 1,2,\cdots,n$. See Figure 3. He assumed that n, ℓ_i, K_i, $i = 1,2,\cdots,n$ are all independent random variables, that the ℓ_i are identically distributed with some exponential distribution, and that the K_i are identically distributed. Aside from assuming all averages of interest are finite, and imposing a technical condition satisfied in "almost all cases", <u>no</u> other assumptions are made about the distributions of the random variables. The regions $x < x_0 = 0$ and $x \geqslant x_n = \sum_{\ell=1}^{n} \ell_i$ are filled by dielectric media of known dielectric constants K_ℓ and K_r.

Now let K_{-p} and ℓ_{-p} be the known dielectric constants and thicknesses of a set of four distinct dielectric slabs, L_p, $p = 1,2,3,4$ which are chosen so that $2\omega^2\mu_0\varepsilon_0 K_{-p}\ell_{-p} = \pi/2$. Further, let K_{+q} and ℓ_{+q} be the known dielectric constants and thicknesses of another set of four distinct dielectric slabs, R_q, $q = 1,2,3,4$ which also satisfy

$2\omega^2\mu_0\varepsilon_0 K_{+q}\ell_{+q} = \pi/2$. Then consider the sixteen cases in which one of the slabs L_p is placed on the left and one of the slabs R_q is placed on the right of the stack of slabs, as in Figure 3. In each case consider a plane electromagnetic wave of frequency ω and electric vector of unit amplitude propagating along the x-axis and incident on the stack of dielectric layers from the left. Let T_{pq} be the amplitude of the transmitted electric field. Gopinath showed that given $\langle 1/|T_{pq}|^2\rangle$, $p,q = 1,2,3,4$, it is possible to calculate $\langle \ell_i \rangle$ and $\langle K_i \rangle$.

In terms of stochastic differential equations, the problem looks like this. In the region $x_0 \leqslant x \leqslant x_n$, the amplitude $e(x)$ of the electric vector is a solution of the differential equation

$$\frac{d^2e(x)}{dx^2} + \beta^2(x,\omega)C(x) = 0. \tag{3.2}$$

In (3.2), $\beta^2(x,\omega)$ is a stochastic process defined by

$$\beta^2(x,\omega) = \omega^2\mu_0\varepsilon_0 K_i, \qquad x_{i-1} \leqslant x < x_i,$$
$$x_0 = 0, \qquad x_i = \sum_{j=1}^{i} \ell_j, \qquad i = 1,2,\cdots,n. \tag{3.3}$$

If $e_0(x)$ and $e_1(x)$ are the two linearly independent solutions of (3.3) satisfying the initial conditions

$$e_0(0) = e_1'(0) = 1, \qquad e_0'(0) = e_1(0) = 0, \qquad (3.4)$$

then the fundamental matrix of (3.2) is

$$G(x) = \begin{bmatrix} e_0(x) & e_1(x) \\ e_0'(x) & e_1'(x) \end{bmatrix}, \qquad (3.5)$$

where $' = \frac{d}{dx}$. It can be shown that $1/|T_{pq}|^2$ is a linear function of the elements of $G^{-1}(x_n) \otimes G^{-1}(x_n)$ where $\otimes$ denotes the tensor product. Thus Gopinath has shown that given $\langle G^{-1}(x_n) \otimes G^{-1}(x_n)\rangle$ it is possible to calculate $\langle \ell_i \rangle$ and $\langle K_i \rangle$. For further details about this remarkable result, the reader should consult Gopinath's paper [1].

4. THE EQUATIONS OF AN ADAPTIVE FILTER. I now want to talk about a completely different part of communications technology where stochastic differential equations appear. Generally, when someone speaks into a telephone, part of the speech signal is reflected back to the speaker from the far end of the circuit. This reflected signal is called an echo and the time it takes the speech signal to get to the far end of the circuit and return to the speaker as an echo is called the round-trip delay. If the echo is large enough and the round-trip delay is long enough, the speaker will find it

difficult to carry on a conversation. With the advent of satellite communications, the problem of controlling echoes has to be looked at again since the round-trip delays are so large. The reader interested in the general problem of echoes in telephone circuits should consult [13].

A new device for controlling echoes, called an "echo canceller", was proposed about a decade ago [14]. Recently Mitra and Sondhi have made a penetrating study of the equations describing its operation. I do not have time to derive the equations here and must refer the reader to their paper for the derivation [2]. The equations are easy to write down, however, and some of their results simple to explain.

The behavior of the echo canceller can be inferred from a knowledge of the "misalignment vector" $\underset{\sim}{r}(t)$, which is a real $n \times 1$ vector valued function of time. In practical situations, n is about 200. The misalignment vector is the solution of the vector differential equation

$$\frac{d}{dt}\underset{\sim}{r}(t) = -KF(\underset{\sim}{r}'(t)\underset{\sim}{x}(t))\underset{\sim}{x}(t). \tag{4.1}$$

Here K is a positive real parameter, $F(\sigma)$ is a real function of a single real variable, $\underset{\sim}{x}(t)$ is a real $n \times 1$ vector valued function of time and $\underset{\sim}{r}'(t)$ is the transpose of $\underset{\sim}{r}(t)$.

The function $F(\sigma)$ is assumed to satisfy the condition

$$\gamma_1\sigma^2 \leqslant F(\sigma) \leqslant \gamma_2\sigma^2 , \tag{4.2}$$

for some γ_1 and γ_2 such that $0 < \gamma_1 < \gamma_2 < \infty$. The special case $F(\sigma) \equiv \sigma$ describes a situation difficult to achieve in practice.

If $x(t)$ is the speech signal of the speaker, then $x_i(t)$, the i-th component of $\underset{\sim}{x}(t)$, is just $x_i(t) = x(t - (i - 1)\Delta)$, $1 \leqslant i \leqslant n$. That is, $x_i(t)$ is just the speech signal delayed by the time $(i - 1)\Delta$, where $\Delta > 0$ is some fixed delay. Since speech is typically considered to be a random process, $\underset{\sim}{x}(t)$ is vector random process and (4.1) is a system of random differential equations. However, Mitra and Sondhi ignored the stochastic nature of $\underset{\sim}{x}(t)$ and imposed the following conditions:

1. There exist constants $\alpha > 0$ and $T > 0$ such that for all $t \geqslant 0$ and all constant vectors $\underset{\sim}{d}$,

$$\frac{1}{T} \int_t^{t+T} [\underset{\sim}{d}'\underset{\sim}{x}(\tau)]^2 d\tau \geqslant \alpha \| \underset{\sim}{d} \|^2 . \tag{4.3}$$

In (4.3), $\|\underset{\sim}{d}\|^2 = \underset{\sim}{d}'\underset{\sim}{d}$ is the square of the Euclidean norm. Condition (4.3) is called a mixing condition.

2. There is a positive number L such that for T as in the mixing condition and for all $t \geqslant 0$,

$$\frac{1}{T} \int_{t}^{t+T} \|\underset{\sim}{x}(\tau)\|^2 d\tau \leq L^2. \tag{4.4}$$

3. $\frac{d}{dt}\underset{\sim}{x}(t)$ exists, and there exist positive constants ℓ and m such that for all $t > 0$, $\|\frac{d}{dt}\underset{\sim}{x}(t)\| \leq \ell$, $\|\underset{\sim}{x}(t)\| \geq m$.

Under these assumptions, Mitra and Sondhi showed that for all $0 \leq t_0 \leq t$,

$$ae^{-b(t-t_0)} \leq \|\underset{\sim}{r}(t)\|/\|\underset{\sim}{r}(t_0)\| \leq ce^{-d(t-t_0)}. \tag{4.5}$$

In (4.5), a, b, c and d are positive constants which depend on K. It is of interest to note that only conditions 1 and 2 were used in obtaining the upper bound, while only condition 3 was used in obtaining the lower bound. Furthermore as $K \to 0$, $b = O(K)$, $d = O(K)$, but as $K \to \infty$, $b = O(\frac{1}{K})$, $d = O(\frac{1}{K})$. This last result is not only remarkable, it is important. This is because the echo canceller is to be designed so that for any initial condition $\underset{\sim}{r}(t_0)$, $\underset{\sim}{r}(t)$ should decay to zero as rapidly as possible. Their result shows that there is an optimum choice for the parameter K. In particular it is not best to make K as large as possible.

The equations discussed here (or very similar ones) arise in a number of different areas of technology. For some references the reader should consult the paper by Sondhi and Mitra. It would thus seem well worthwhile to study them further and try to make direct use of their stochastic nature.

5. STOCHASTIC DIFFERENCE EQUATIONS.

Finally, I want to briefly mention the subject of stochastic difference equations. Devices called "tapped delay lines" play a prominent role in signal processing. In recent years, a new device called a "Charge Coupled Device" (CCD) has been developed, and one of its important applications is the implementation of delay lines [15]. Because of space limitations, I will not discuss CCD's here, but must refer the reader to [15] which is an excellent review of the subject and contains a large bibliography.

The behavior of a tapped delay line when used in signal processing is usually described by a set of difference equations. A typical, important set of such equations describing an ideal tapped delay line is

$$\begin{aligned} x_1(k+1) &= a_1 x_1(k) + a_2 x_2(k) + u(k), \\ x_2(k+1) &= x_1(k), \end{aligned} \tag{5.1}$$

for $k = 0,1,2,\cdots$. In (5.1), a_1 and a_2 are constants, and $u(k)$ is a given sequence. However, when the tapped delay line is implemented with CCD's, Equations (5.1) must be modified to take into account a property of CCD's called charge transfer inefficiency. When this is taken into account, Equations (5.1) become [16]

$$x_1(k+1) = \varepsilon_1 x_1(k) + a_1(1-\varepsilon_1)x_1(k) + a_2(1-\varepsilon_2)x_2(k) + u(k),$$

$$x_2(k+1) = (1-\varepsilon_1)x_1(k) + \varepsilon_2 x_2(k). \tag{5.2}$$

In (5.2), ε_1 and ε_2 are random variables, and in some cases it may be necessary to assume that $\varepsilon_1 = \varepsilon_1(k)$ and $\varepsilon_2 = \varepsilon_2(k)$ are sequences of random variables. In all cases of interest, it is true that $0 \leqslant \varepsilon_i \ll 1$, $i = 1,2$.

The theory of stochastic difference equations is probably not too different from the theory of stochastic differential equations, but I am sure it will have applications in signal processing and is worth investigating.

REFERENCES

1. B. Gopinath, *The Solution to an Inverse Problem in Stratified Dielectric Media*, J. Math. Phys., to be published.
2. M. Sondhi and D. Mitra, *New Results on the Performance of a Well-Known Class of Adaptive Filters*, Proc. IEEE, to be published.
3. W. D. Warters, *Millimeter Waveguide Scores High in Field Test*, Bell Laboratories Record, 53 (November 1975), 400-408.
4. T. A. Abele, D. A. Alsberg and P. T. Hutchinson, *A High-Capacity Digital Communications System Using TE_{01} Transmission in Circular Waveguide*, IEEE Trans. Microwave Theory and Techniques, MTT-23 (April 1975), 326-333.
5. S. A. Schelkunoff, *Generalized Telegraphist's Equations for Waveguides*, B.S.T.J., 31 (July 1952), 784-801.
6. W. H. Louisell, *Coupled Mode and Parametric Electronics*, New York: John Wiley, 1960.
7. W. K. H. Panofsky and M. Phillips, *Classical Electricity and Magnetism*, New York: Addison-Wesley, 1955, Chap. 12.
8. S. P. Morgan, *Theory of Curved Circular Waveguide Containing an Inhomogeneous Dielectric*, B.S.T.J., 36 (September

1957), 1209-1251.

9. H. E. Rowe and W. D. Warters, *Transmission in Multimode Waveguide with Random Imperfection*, B.S.T.J., 41 (May 1962), 1031-1170.

10. J. A. Morrison and J. McKenna, *Coupled Line Equations With Random Coupling*, B.S.T.J., 51 (January 1972), 209-228.

11. J. Millman and H. Taub, *Pulse and Digital Circuits*, 1st Edition, New York: McGraw Hill, 1956, 28-31.

12. B. S. T. J., 50 (April 1971).

13. J. W. Emling and D. Mitchell, *The Effects of Time Delay and Echoes on Telephone Conversations*, B. S. T. J., 42 (November 1963), 2869-2892.

14. M. M. Sondhi, *An Adaptive Echo Canceller*, B. S. T. J., 46 (March 1967), 497-511.

15. C. H. Séquin and M. F. Tompsett, *Charge Transfer Devices*, New York: Academic Press, 1975.

16. B. Gopinath and A. Gersho, *Multiplexes Filtering With Charge Transfer Devices*, IEEE J. Solid State Circuits, SC-11 (February 1976), 220-224.

Bell Laboratories
Murray Hill, N. J.

Diffusion on the Line and Additive Functionals of Brownian Motion

BY

Steven Orey

1. Diffusion on the Line.

One of the major achievements of modern probability theory is to give us quite explicitly all diffusion processes with stationary transition probabilities on a linear interval I. Here we mean by a *diffusion* a strong Markov process with continuous sample paths. The problem is solved in several steps. First attention is restricted to processes X_t with infinite life times, i.e. $P[X_t \in I] = 1$ for all $t \geqslant 0$, and with every point x in the interior of I assumed to be *regular* in the sense that for any point y in the interior of I, the diffusion started at x will hit y at some time, with probability one. To study such diffusions Feller first introduced a scale function $p(x)$ (i.e. a continuous strictly increasing function) on I so that relative to this scale the process becomes fair, which is to say that $\overline{X}_t = p(X_t)$ has the property that for $[x - \delta, x + \delta]$ included in the interior of $p(I)$, the process $\overline{X}$ started at x will get out of $[x - \delta, x + \delta]$ with probability one, and the probability that it exits at $x + \delta$ is one half. Thus $\overline{X}_t$ acts just like Brownian motion as long as it is in the interior of $p(I)$, *except* that it may move at a nonconstant rate, depending on its position. The second, and more difficult step taken by Feller, was to introduce a measure $m(dx)$ on the interior of I to gauge the speed with which $p(X_t)$ is moving when $X_t = x$. This measure is called the *speed measure*. It is finite on compact intervals included in the interior of I, and assigns strictly positive mass to open intervals. The scale p and speed measure m determine X_t up to the first time T that X_t exits the interior of I. In case $T < \infty$ with positive probability the further motion can be completely described by suitably defining m on the end points. What makes the solution complete is the fact that every choice of p and m leads to a unique diffusion; see Ito-McKean [4], Dynkin [1]. The theory has been extended to allow killing (i.e. finite life times) and points that are not regular; see [4]. A restriction such as requiring all point to be regular is a natural probabilistic restriction; by contrast restrictions which require that the generator have some pre-assigned analytic form do not appear germaine.

On the other hand the problem of finding the general diffusion process in R^n (even with stationary transition probabilities) seems to be hopelessly difficult. The most one can do now is to give broad classes of differential operators which correspond to diffusions. Among the best results at present are those of Stroock-Varadhan [6].

This work was motivated by the idea that the next situation in which we might hope to know the general diffusion is that in which the process again moves on a linear interval I, but stationary transition probabilities are not assumed. We will assume there is no killing. Furthermore we are quite ready to assume "regularity" of all points in a suitable sense. Finally our interest will focus on the behavior of the diffusion inside I, up to the time of reaching the boundary only.

Consider specifically the situation in which Z'_t is a continuous strong Markov process on I. Volkonskii [5] introduced a suitable notion of a regular point, and showed that if all points in the interior of I are regular then one can introduce a function $p(x,t)$ continuous for $(x,t) \in I \times [0,\infty]$, increasing in x, and such that $Z_t = p(Z'_t,t)$ is fair as long as Z_t remains in the interior of I.

The problem of finding the general diffusion on R_1 is therefore closely linked to that of finding all those diffusions on R_1 which are martingales, and this will be our primary problem here. Note that Z_t being a diffusion is equivalent to (Z_t,t) being a diffusion with stationary transition probabilities. Also, since Z_t is a continuous martingale there is associated an increasing continuous process τ_t, with $\tau_0 = 0$, namely the square variation process (often denoted by $[Z,Z]_t$, see [3]), and if $\phi_t = \inf\{s: \tau_s \geq t\}$ then Z_{ϕ_t} is Brownian motion. To be more precise, let us assume for convenience that $\tau_t \to \infty$ with t, so that ϕ_t is defined for all t. Let $G_t = \sigma(Z_s: s \leq t)$ be the σ-field generated by the Z_s, $0 \leq s \leq t$. Then

i) (Z_{ϕ_t}, ϕ_t) is a diffusion with stationary transition probabilities,

ii) $(Z_{\phi_t}, G_{\phi_t}, \ 0 \leq t < \omega)$ is a Brownian motion martingale.

Henceforth $X_t = Z_{\phi_t}$, and $F_t = \sigma(X_s: 0 \leq s \leq t)$. Note that $F_t \subseteq G_{\phi_t}$.

We are lead then to consider the following problem. Given a Brownian motion X_t, with $F_t = \sigma(X_s: s \leq t)$, find all processes ϕ_t, defined on a possibly larger probability space, such that ϕ_t is continuous, increasing, $\phi_0 = 0$ and (X_t, ϕ_t) is a diffusion process with homogeneous transition probabilities. Such ϕ_t will be called *continuous increasing randomized additive functionals*, and the class will be denoted by Φ. A process ϕ is *strictly increasing* if with probability one $t \to \phi_t(\omega)$ is strictly increasing everywhere. Φ^+ will denote the strictly increasing members of Φ. The word "randomized" is used because we have not demanded that

ϕ_t be measurable with respect to F_t. Roughly the continuous increasing additive functionals should be those $\phi_t \in \Phi$ defined on the original (unenlarged) probability space and such that ϕ_t is F_t measurable for each t. Indeed we suspect that this restricted class is in fact no smaller than Φ. Essentially the same problem: is the inclusion $F_t \subseteq G_{\phi_t}$ noted at the end of the last paragraph ever proper?

For $\phi \in \Phi$, define $\tau_t = \inf\{s: \phi_s > t\}$. Then $Z_t = X_{\tau_t}$ is a strong Markov process; if $\phi \in \Phi^+$, Z_t will be a diffusion, that is, it will have continuous sample paths. The transition probabilities of (Z_t) will be stationary only if ϕ is a homogeneous additive functional of Brownian motion. This class has been extensively studied and is completely known. In this case one obtains that ϕ_t is F_t-measurable; for further discussion of this case see Section 4.

2. Additive Functions.

It will be convenient to have the underlying Brownian motion X presented as a Markov process with stationary transition probabilities:

$$X = (\Omega, F, F_t, X_t, \Theta_t, P^x),$$

see for example [1], p. 20. For F_t we take $\cap_{\varepsilon>0} F^0_{t+\varepsilon}$, where $F^0_t = \sigma(X_s: 0 \leq s \leq t)$; $F = \vee_{t<\omega} F_t$. The Θ_t are the shift operators and P^x is the probability on (Ω, F) corresponding to the process X started at x.

As already indicated, the continuous, increasing additive functionals of Brownian motion we seek are essentially the elements of Φ which are F_t-adapted. To motivate our definition, we consider a typical example.

EXAMPLE: Let $g(x,t)$ be a non-negative continuous function on $(-\infty,\infty) \times [0,\infty]$, and suppose g is Lipschitz continuous in t. The differential equation $d\phi_t = g(X_t,t)dt$, $\phi_0 = 0$ has, with probability one, a unique solution. Note that $\phi \in \Phi$ and ϕ_t is F_t-measurable. For our purpose it is natural also to consider the differential equation with initial condition $\phi_0 = s$, $s > 0$; denote the unique solution by ϕ^s. The collection (ϕ^s), $0 \leqslant s < \infty$, is the paradigm of what we will call a continuous, increasing, additive functional.

DEFINITION. The collection (ϕ^s), $0 \leqslant s < \infty$, belongs to Φ_0 if and only if each ϕ^s is a continuous, increasing stochastic process satisfying the following conditions:

i) ϕ_t^s is measurable with respect to F_t, $0 \leqslant t < \infty$,

ii) $\phi_0^s = s$, P^x - a.s. for every $x \in (-\infty,\infty)$,

iii) For every stopping time β, $s \geqslant 0$, $t \geqslant 0$, $x \in (-\infty,\infty)$

$$\phi_{\beta+t}^s(\omega) = \phi_t^{\alpha(\omega)}(\Theta_\beta\omega) \quad P^x \text{ - a.s.} \qquad (*)$$

where $\alpha(\omega) = \phi_{\beta(\omega)}^s(\omega)$, and of course β means $\beta(\omega)$ at both occurrences in (*).

DEFINITION. Φ_0^+ is the class of $(\phi^s) \in \Phi$ such that each ϕ^s is a strictly increasing process.

Examples will be given at the end of this section. We turn to some basic properties. A familiar formulation of the strong Markov property is that for any bounded random variable γ measurable with respect to F and any stopping time β

$$E^x[\gamma \circ \Theta_\beta | F_\beta] = E^{X_\beta}[\gamma];$$

see [1]. We require a slight extension.

LEMMA. Let $\gamma(s,\omega)$ be a bounded function on $[0,\infty] \times \Omega$ measurable with respect to $R \times F$, where R is the class of Borel sets on $[0,\infty]$. Let β be a stopping time and α an F_β-measurable random variable. Set $q(x,s) = E^x[\gamma(s,\cdot)]$. Then for $x \in (-\infty,\infty)$,

$$E^x[\gamma(\alpha,\Theta_\beta)|F_\beta] = q(X_\beta,\alpha).$$

PROOF. When $\gamma(s,\omega)$ factors into $\gamma_1(s)\gamma_2(\omega)$ the result follows from the usual form of the strong Markov property. By linearity it extends to linear combinations of such products, and finally the general case follows by a monotone class argument.

PROPOSITION 1. Let $(\phi^s) \in \Phi_0$. For $u \geq 0$, $(X_t, \phi_t^u, \ 0 \leq t < \infty)$ is a strong Markov process on $(-\infty,\infty) \times [u,\infty)$ with homogeneous transition probabilities

$p_t((x,s),D) = P^x[(X_t,\phi_t^s) \in D]$, where $(x,s) \in (-\infty,\infty) \times (u,\infty)$ and D is a measurable subset of $(-\infty,\infty) \times [u,\infty)$.

PROOF. Let $g(x_1,y_1,\cdots,x_m,y_m)$ be a bounded Borel function, the x_i ranging over $(-\infty,\infty)$, the y_i over $[u,\infty]$. Let $t_1,t_2,\cdots,t_m$ be non-negative numbers and set

$$q(x,s) = E^x g(X_{t_1},\phi_{t_1}^s,\cdots,X_{t_m},\phi_{t_m}^s).$$

Then for any stopping time β

$$E[g(X_{\beta+t_1},\phi_{\beta+t_1}^u,\cdots,X_{\beta+t_m},\phi_{\beta+t_m}^u)|F_\beta] = q(X_\beta,\phi_\beta^u),$$

since, on setting $\alpha = \phi_\beta^u$ the left side can be written

$$E[g(X_{t_1}(\Theta_\beta),\phi_{t_1}^\alpha(\Theta_\beta),\cdots,X_{t_m}(\Theta_\beta),\phi_{t_m}^\alpha(\Theta_\beta))|F_\beta]$$

and the lemma applies.

For $(\phi^s) \in \Phi_0$, it follows from Proposition 1 that $\phi^0 \in \Phi$. Another very important property of (ϕ^s) is given by the next proposition.

PROPOSITION 2. Let $(\phi^s) \in \Phi_0$. For $0 \leq s \leq s'$, $x \in (-\infty,\infty)$,

$$P^x[\phi_t^s \leq \phi_t^{s'}, \quad 0 \leq t < \infty] = 1.$$

PROOF. Let $\beta = \inf\{t: \phi_t^s = \phi_t^{s'}\}$, with $\beta = \infty$ if not otherwise defined. Obviously $\phi_t^s < \phi_t^{s'}$ for all t on $\beta = \infty$. On $\beta < \infty$ define α to be the common value of ϕ_β^s and $\phi_\beta^{s'}$ and use the defining property of Φ_0 to obtain

$$\phi^{s}_{\beta+t} = \phi^{\alpha}_{t}\circ\Theta_{\beta} = \phi^{s'}_{t}$$

P^x - a.s. The equalities hold P^x - a.s. for each fixed t, but since the functionals are continuous, one obtains P^x - a.s. equality for all t.

We turn to some classes of examples.

A) HOMOGENEOUS ADDITIVE FUNCTIONALS. These are members of Φ_0 satisfying $\phi^s_t = s + \phi^0_t$. It is known that they can be represented

$$\phi^0_t = -e(X_t) + e(X_0) + \int_0^t e'(X_s)dX_s, \qquad (1)$$

where e is a convex function, e' the almost everywhere defined derivative of e, and the second derivative of e exists as a measure m, finite on compact sets, and one has the alternative representation

$$\phi^0_t = \int L(t,x)m(dx), \qquad (2)$$

where the random variable $L(t,x)$ is the local time of the Brownian motion at position x, time t. The representation (1) can be deduced from a result of Tanaka; (2) is due to Ito and McKean [4]. For an elegant derivation see Wang [7]. If J is a compact interval, and $e_J(x)$ is the expected value of the first exit time of Brownian motion started at x from J, then $e_J(x)$ is intimately related to $e(x)$; in fact the second derivative of e_J agrees with that of

e on J.

B) Let $(\phi^s) \in \Phi_0$, (or Φ_0^+) and let g be a continuous, strictly increasing function on $[0,\infty)$, vanishing at 0, with inverse g^{-1}. Setting $\psi_t^s = g(\phi_t^{g^{-1}(s)})$, one finds that again $(\psi^s) \in \Phi_0$ (respectively Φ_0^+). Special cases are

b 1) $\phi_t^0 = t$, giving $\psi_t^0 = g(t)$;

b 2) (ϕ^s) a homogeneous additive functional, $g(t) = e^t - 1$, giving (ψ^s) such that $\psi_t^s = \psi_t^0(1 + s) + s$.

C) Suppose (ψ^s) is a homogeneous additive functional and $g(x,t)$ is a positive function sufficiently smooth so that for each $s > 0$ and every Brownian path $t \to X_t$ the equation

$$\phi_t^s = s + \int_0^t g(X_u, \phi_u^s) d\psi_u^0$$

has a unique solution. Then $(\phi^s) \in \Phi_0$.

3. The Potential.

Let (a,b) be a finite open interval. Henceforth

$$T = \inf\{t: X_t \notin (a,b)\}.$$

Given $\Psi \in \Phi$, define

$$e(x,s) = E[\Psi_T - \Psi_0 | X_0 = x, \Psi_0 = s], \quad -\infty < x < \infty, \quad 0 \leqslant s < \infty.$$

For $x \notin (a,b)$, $e(x,s) = 0$. We will be interested in the case in which $e(x,s) < \infty$ for all x and s. This restriction is not a serious one. Call e the *potential* of Ψ. If $(\phi^s) \in \Phi_0$ the corresponding potential can be defined by

$$e(x,s) = E^x[\phi^s_T - s].$$

We introduce the notation

$$e^*(x,s) = e(x,s) + s.$$

THEOREM 1. Let $\psi \in \Phi$ have finite potential e. Then

$$e^*(x,s) \text{ is non-decreasing in } s \text{ for fixed } x \qquad \text{(i)}$$

provided $\psi \in \Phi_0$, i.e. $\psi = \phi^0$ for some $(\phi^s) \in \Phi_0$. Furthermore if (i) then also

$$e(x,s) \text{ is convex in } x \text{ for fixed } s. \qquad \text{(ii)}$$

PROOF. If $\psi \in \Phi_0$, Proposition 2 implies that for $s' > s$

$$e(x,s') - e(x,s) = E[\phi^{s'}_T - \phi^s_T] - (s' - s) \geqslant -(s' - s)$$

proving (i).

Now let $\psi \in \Phi$ and assume (i) holds. Let $a < a' < b' < b$, $x = \frac{1}{2}(a' + b')$ and let $T_1 = \inf\{t: X_t \notin (a',b')\}$. Then

$$e(x,s) = E[(\psi_T - \psi_{T'}) + (\psi_{T'} - \psi_0) | (X_0,\psi_0) = (x,s)]$$

$$= \int_{s'>s} (e(a's') + (s'-s)) P[X_{T_1} = a', \psi_{T_1} \in ds' | X_0 = x, \psi_0 = s]$$

$$+ \int_{s'>s} (e(b',s') + (s'-s)) P[X_{T_1} = b', \psi_{T_1} \in ds' | X_0 = x, \psi_0 = s]$$

$$\geqslant \int_{s'>s} e(a',s) P[X_{T_1} = a', \psi_{T_1} \in ds' | X_0 = x, \quad \psi_0 = s]$$

$$+ \int_{s'>s} e(b',s) P[X_{T_1} = b', \psi_{T_1} \in ds' | X_0 = x, \quad \psi_0 = s]$$

$$= \frac{1}{2}(e(a',s) + e(b',s)).$$

As long as we are interested in X_t only for $0 \leqslant t \leqslant T$ it is convenient to define $\overline{X}_t = X_{t\wedge T}$, and, for $\psi \in \Phi$, to consider $\overline{\psi}_t = \psi_{t\wedge T}$. To avoid clumsy notation we omit the bar, write simply X_t, ψ_t for the stopped processes.

PROPOSITION 3. If $\psi \in \Phi$ has a finite potential e, $e^*(X_t,\psi_t)$ is a martingale, with respect to $P = P^x$, $a \leqslant x \leqslant b$.

PROOF. If $H_t = \sigma((X_u,\psi_u): \ 0 \leqslant u \leqslant t)$, then

$$e(X_t,\psi_t) = E[\psi_\infty - \Psi_t | H_t],$$

and so

$$e^*(X_t,\psi_t) = E[\psi_\infty | H_t].$$

Let $e(x,t)$ be a function on $[a,b] \times [0,\infty)$, with two continuous derivatives in x, one continuous derivative in t, and let ϕ_t be a continuous stochastic process with sample functions of bounded variation. By Ito's formula

$$d\phi_t + de(X_t,\phi_t) = (1 + \frac{\partial e}{\partial t}(X_t,\phi_t))d\phi_t + \frac{1}{2}\frac{\partial^2 e}{\partial x^2}(X_t,\phi_t)dt + \frac{\partial e}{\partial x}(X_t,\phi_t)dX_t. \qquad (1)$$

If now $\phi \in \Phi$, with potential e, Proposition 3 asserts that the left side of (1) is the differential of a continuous martingale; and the same is true of the last term in (1). Hence the sum of the first two terms on the right side of (1) corresponds to a continuous martingale of bounded variation. In fact then this martingale must vanish, giving

$$\frac{d\phi}{dt} = \frac{-\frac{1}{2}\frac{\partial^2 e}{\partial x^2}(X_t,\phi_t)}{1 + \frac{\partial e}{\partial t}(X_t,\phi_t)} \equiv \frac{1}{a(X_t,\phi_t)}, \qquad (2)$$

hence also

$$d\phi_t - de(X_t,\phi_t) = \frac{\partial e}{\partial x}(X_t,\phi_t)dX_t. \qquad (3)$$

REMARK. Let $e(x,s)$ be a non-negative function belonging to $C^{2,1}([a,b] \times [0,\omega))$ and vanishing for $x \in \{a,b\}$. Suppose the corresponding $a(x,s)$ defined in (2) is continuous, positive and such that $(a(x,t))^{-1}$ is Lipschitz continuous in t. Then (2) will have a unique solution and so *there exists at most one* $\phi \in \Phi$ *with poten-*

tial e, *and such a* $\phi \in \Phi_0$; furthermore, *if* $\partial e/\partial x$ *is bounded the solution* ϕ *of* (2) *will satisfy* $\phi \in \Phi_0$ *and* e *will be the potential* of ϕ. The assumption $\partial e/\partial x$ bounded is made to justify application of the optional sampling theorem in (1), to conclude that e is the potential of ϕ.

Let now $(\phi^s) \in \Phi_0$ with finite potential $e(x,s)$. Fix s_0 and write ϕ_t for $\phi_t^{s_0}$. Then

$$Y_t = e^*(X_t, \phi_t)$$

is a martingale adapted to (F_t), and so it has a representation

$$Y_t = Y_0 + \int_0^t H_s dX_s, \qquad (4)$$

where (H_s) is adapted to (F_s), $\int_0^t H_s^2 ds < \infty$ a.s.. Recall that $e^*(x,t)$ is non-decreasing in t, convex in x. Let

$$D_e = \{(x,t)\colon a < x < b,\ 0 < t < \infty,\ \frac{\partial e}{\partial x}(x -, t) > \frac{\partial e}{\partial x}(x +, t)\}.$$

The set D_e is a Lebesgue null set (each t-section being in fact denumerable). I do not know whether $P^x[(X_t, \phi_t) \in D_e] = 0$. Let

$$D_e^* = \{x\colon a < x < b \text{ and } \frac{\partial e}{\partial x}(x -, t) > \frac{\partial e}{\partial x}(x +, t) \text{ for some } t \in (0, \infty]\}.$$

We shall write

$$e'(x,t) = \frac{\partial e}{\partial x}(x+,t).$$

THEOREM 2. Assume D_e^* is a Lebesgue null set. Then

$$Y_t - Y_0 = (\phi_t + e(X_t,\phi_t)) - (\phi_0 + e(X_0,\phi_0)) = \int_0^t e'(X_s,\phi_s)dX_s. \tag{5}$$

PROOF. According to a theorem of Isaacson [2], there exists a Lebesgue null set S_0 on the time axis such that the integrand H_s in (4) can be obtained as

$$H_s = \text{P-}\lim_{t \downarrow s} \frac{Y_t - Y_s}{X_t - X_s}, \quad s \in [0,T)\setminus S_0,$$

where the limit is a limit in probability. Now

$$\frac{Y_t - Y_s}{X_t - X_s} = \frac{e^*(X_t,\phi_t) - e^*(X_t,\phi_s)}{X_t - X_s} - \frac{e^*(X_t,\phi_s) - e^*(X_s,\phi_s)}{X_t - X_s}$$

and for $X_s \notin D_e^*$ (an event having probability one) the last term tends to $e'(X_s,\phi_s)$ as $t \downarrow s$ a.s. So for small $t - s$, the conditional probability given (X_s,ϕ_s) that the first term on the right is near $H_s - e'(X_s,\phi_s)$ is near one. However this first term has the same sign as $X_t - X_s$ (by (i) of Theorem 1) and since

$$P[(X_t - X_s) > 0 | X_s,\phi_s] = P[(X_t - X_s) < 0 | X_s,\phi_s] = 1/2,$$

we conclude $H_s - e'(X_s,\phi_s) = 0$, as desired.

Suppose $e^*(x,\cdot)$ is strictly increasing, for each x.

Then we can introduce the inverse function $q(x,\cdot)$, and define

$$\tilde{e}(x,y) = e'(x,q(x,y)).$$

Observe that (5) then becomes

$$Y_t - Y_0 = \int_0^t \tilde{e}(X_t,Y_t)dX_s \tag{6}$$

and $\phi_t = q(X_t,Y_t)$.

PROPOSITION 4. Let $e^*(x,s) = e(x,s) + s$ be a continuous non-negative function on $(a,b) \times [0,\infty)$. Suppose e^* is convex in x and strictly increasing in s. If there exists a finite constant c such that

$$|e'(x,t) - e'(x,s)| < c|e^*(x,t) - e^*(x,s)|,$$

$$x \in (a,b), \quad 0 \leqslant s \leqslant t, \tag{7}$$

then for given $s_0 \geqslant 0$ there exists a unique solution ϕ_t of (5) satisfying $\phi_0 = s_0$, with t ranging over $[0,T)$.

PROOF. Condition (7) is equivalent to $\tilde{e}(x,y)$ satisfying a Lipschitz condition in y, with constant c. So by Ito's theory (6) has a unique solution, subject to $Y_0 = y_0 = e^*(x_0,s_0)$, say, and then $\phi_t = q(X_t,Y_t)$ is the desired solution to (5). (Uniqueness here means that for any other solution $\tilde{\phi}_t$, $\tilde{\phi}_t = \phi_t$ for all t P^{x_0} - a.s.)

PROPOSITION 5. In addition to the assumptions of Proposition 4, assume that D_e^* has Lebesgue measure zero, and $e'(x,t)$ is bounded on $(a,b) \times [0,t_0]$ for all $t_0 \geq 0$, and continuous in x and t on the complement of $D_e^* \times [0,\infty)$. Then the solution ϕ_t of (5) is increasing.

PROOF. Suppose first that $e(x,t)$ is three times continuously differentiable. Since $e^*(x,q(x,y)) = y$, we obtain

$$\frac{\partial q}{\partial x}(x,y) = -\frac{\partial e^*}{\partial x}(x,q(x,y)) \cdot \left(\frac{\partial e^*}{\partial y}(x,q(x,y))\right)^{-1},$$

$$\frac{\partial q}{\partial y}(x,y) = \left(\frac{\partial e^*}{\partial y}(x,q(x,y))\right)^{-1},$$

and applying Ito's formula to $\phi_t = q(X_t,Y_t)$ one obtains, if one also remembers (6), that $\phi_t - \phi_0$ is represented as an integral with respect to t only. Thus ϕ_t is indeed absolutely continuous. So again we have the formula (1). Evidently, from (5), $\phi_t + e(X_t,\phi_t)$ is a continuous martingale: so as before, (2) must hold. The assumptions on e insure that $a(x,t)$ is positive, and so ϕ_t is increasing in this case. Note $Y_t = e^*(X_t,\phi_t) \geq \phi_t$.

If e is not smooth, approximate e by a sequence e_n of three time continuously differentiable functions satisfying the same hypotheses as e. For $t_0 > 0$, $\varepsilon > 0$ fixed, we require that e_n converges to e uniformly on $a \leq x \leq b$, $0 \leq t \leq t_0$, and that $\partial e_n/\partial x$ converges to $\partial e/\partial x$ on the compliment of D_e^*, the convergence being uniform on $[a,b]\setminus D^\varepsilon \times [0,t_0]$, where D^ε is an open set including D_e^* of measure at most ε. For each e_n, there is a $\phi^{(n)}$ satisfying the corresponding differential equation. This means $Y_t^{(n)} = e^*(X_t, \phi_t^{(n)})$ satisfies

$$Y_t^{(n)} = Y_0^{(n)} + \int_0^t \tilde{e}(X_s, Y_s^{(n)}) dX_s.$$

It is not evident that $\phi_t^{(n)}$ converges to ϕ_t. For $t_1 > 0$, let μ_n be the measure induced by the process $(X_s, Y_s^{(n)})$, $0 \leq s \leq t_1$, on the space of continuous functions from $[0, t_1]$ onto R^2. A weak convergence argument (see Stroock-Varadhan [6]) shows that there is a subsequence, μ_n, converging weakly to some limit μ. Because (6) has a unique solution Y_t, μ is identified as the measure induced by (X_t, Y_t). It follows that $\phi_t = q(X_t, Y_t)$ inherits the property of being non-decreasing from the $\phi_t^{(n)}$.

THEOREM 3. Let $e^*(x,s)$ be a continuous non-negative function on $[a,b] \times [0,\infty)$ vanishing for $x = a$ and $x = b$. Suppose e^* is convex in x and strictly increasing in s. Assume the Lipschitz condition (7) holds. Assume that D_e^* has Lebesgue measure zero, e' is bounded, and continuous for x in the complement of D_e^*. Then there exists a unique $(\phi^s) \in \Phi_0$ having e as potential.

PROOF. For each fixed s, there is a unique solutio[n] ϕ^s of (5) with $\phi_0^s = s$, by Proposition 4. By Proposition 5, ϕ^s is non-decreasing. The other properties in the definition of Φ_0 are easily verified. To verify that e is the potential of (ϕ^s) proceed as in the Remark following (3). Theorem 2, together with the uniqueness assertion in Proposition 4, shows that there can be only one $(\phi^s) \in \Phi_0$ with potential e, under our assumptions.

4. STATIONARY TRANSITION PROBABILITIES.

Suppose now that our diffusion martingale Z_t has stationary transition probabilities. In that case it is known that ϕ_t is a homogeneous additive funtional, and in particular ϕ_t is F_t-measurable. We give an easy proof within the above framework. Indeed then

$$e(x,s) = e(x) = E_x[S],$$

where $S = \inf\{t: Z_t \notin (a,b)\}$. So evidently $e^*(x)$ is increasing in x, and by (ii) e is convex. By Proposition 3, $(e(X_t) + \phi_t, G_t)$ is a martingale. Equation (5), with initial condition $\tilde{\Phi}_0 = 0$ now becomes

$$\tilde{\Phi}_t = \int_0^t e'(X_s)dX_s - e(X_t) + e(X_0)$$

and of course this has a unique "solution", measurable with respect to F_t. But, $e(X_t) + \tilde{\Phi}_t$ is also a continuous martingale with respect to G_{ϕ_t}. Also $\tilde{\Phi}_t$ is increasing (e.g. by Proposition 5). So $(\phi_t - \tilde{\Phi}_t)$ is a continuous martingale with respect to (G_t), and of bounded variation. So $\phi_t = \tilde{\Phi}_t$ as desired.

REFERENCES

1. E. B. Dynkin, *Markov Processes*, Academic Press, New York (1965).

2. D. Isaacson, *Stochastic Integrals and Derivatives*, Ann. Math. Stat. 40 (1969), 1610-1616.

3. P. A. Meyer, *Integrales Stochastiques*, Springer Lecture Notes in Math. 39 (1967).

4. K. Ito and H. P. McKean, *Diffusion Processes and Their Sample Paths*, Academic Press, New York (1965).

5. V. A. Volkinskii, *Construction of Inhomogeneous Markov Processes by Means of a Random Time Substitution*, Teor. Veroyatnost. i ee Priminen 6 (1961), 47-56.

6. D. W. Stroock and S. R. S. Varadhan, *Diffusion Processes with Continuous Coefficients*, Comm. Pure & Appl. Math. 22 (1969), 345-400, 479-530.

7. A. Wang, *Quadratic Variation of Functional of Brownian Motion*, Tech. Report, Dept. Math., Univ. of Tenn. (1975).

An Individual Ergodic Theorem for the Diffusion on a Manifold of Negative Curvature

BY

Mark A. Pinsky

1. Introduction.

The purpose of this note is to extend some of the recent methods and results [1-5] on the asymptotic behavior of $r(t)$, the geodesic distance from a fixed origin for the Brownian motion on a simply connected, complete, 2-dimensional Riemannian manifold of curvature $K \leq 0$.

The main result states that a.s.

$$\lim_{t\to\infty} \frac{r(t)}{t} = \left\{ \lim_{r\to\infty} \frac{\int_0^r -KG\,dr}{\int_0^r G\,dr} \right\}^{1/2}, \tag{1.1}$$

where G is the density of the Riemannian volume element. The existence of the right member implies the existence of the left member. It is an open problem to find necessary and sufficient conditions for the existence of $\lim_{t\to\infty} r(t)/t$.

In a final section we give a new result on the spectrum of the Laplacian.

2. Preliminary Curvature Estimates.

Let M be a simply connected, 2-dimensional, complete Riemannian manifold of a non-positive curvature. In a system of geodesic polar coordinates about $0 \in M$, we can write

$$ds^2 = dr^2 + G^2(r,\theta)d\theta^2, \tag{2.1}$$

where $G(0,\theta) = 0, G_r(0,\theta) = 1$ and $-G_{rr}/G = K$, the Gaussian curvature. By the theorem of Cartan-Hadamard we may suppose that M is R^2 with the Riemannian metric (2.1). The following estimate is recorded for completeness.

LEMMA 2.1. $G_r/G \geqslant 1/r$ *for* $r > 0$.

PROOF. Let $\psi = G/G_r$. Then $\psi(0) = 0$ and $\psi = (G_r^2 - G\,G_{rr})/G_r^2 \leqslant 1$. Thus $\psi(r) \leqslant r$.

In particular G_r/G is bounded away from zero on any compact subset of $[0,\infty)$.

In the following two lemmas we will show that

$$\liminf_{r\to\infty} \sqrt{-K} \leqslant \liminf_{r\to\infty} G_r/G \leqslant \limsup_{r\to\infty} G_r/G \leqslant \limsup_{r\to\infty} \sqrt{-K}.$$

LEMMA 2.2. *If* $\liminf_{r\to\infty} G_{rr}/G \geqslant k^2 > 0$, *then* $\liminf_{r\to\infty} G_r/G \geqslant |k|$.

PROOF. We may suppose without loss of generality that $k = 1$. Indeed, it suffices to define $\tilde{G}(r,\theta) = k\,G(r/k,\theta)$. Then $\tilde{G}_{rr}/\tilde{G} = k^{-2}G_{rr}/G$ and $\tilde{G}_r/\tilde{G} = k^{-1}G_r/G$.

For the proof, we let $h = G_r/G$; we must show that

$\liminf_{r\to\infty} h \geq 1$.

We consider two cases.

If $\limsup_{r\to\infty} h(r) = \liminf_{r\to\infty} h(r)$, let m be the common value. If $m < 1$, let $\varepsilon < (1 - m)/2$. For all sufficiently large r, we have $G_{rr}/G \geq (1 - \varepsilon)^2$. Now, from the differential equation for h we have $h_r = G_{rr}/G - h^2 \geq (1-\varepsilon)^2 - (m+\varepsilon)^2 > 0$ for all sufficiently large r. Thus $h(r) > c_1 + c_2 r$ for some positive c_2 and thus $\liminf_{r\to\infty} h(r) = \infty$, a contradiction. Thus $m \geq 1$.

In the second case we may assume that $\limsup_{r\to\infty} h(r) > \liminf_{r\to\infty} h(r)$. If the latter is strictly less than 1, we may construct an infinite sequence $r_k \to \infty$ with $h(r_k) = A$, $h_r(r_k) \leq 0$ with $A < 1$. Choose $\varepsilon < 1 - A$; for all sufficiently large r, we have $G_{rr}/G \geq (1 - \varepsilon)^2$. Now, from the differential equation we have $h_r(r_k) = G_{rr}/G - h^2(r_k) \geq (1-\varepsilon)^2 - A^2 > 0$, a contradiction. Therefore $\liminf_{r\to\infty} h(r) \geq 1$, which was to be proved.

LEMMA 2.3. *If* $\limsup_{r\to\infty} G_{rr}/G \leq k^2$, *then* $\limsup_{r\to\infty} G_r/G \leq |k|$.

PROOF. As before, we may assume that $k = 1$. Defining $h = G_r/G$, we must show that $\limsup_{r\to\infty} h(r) \leq 1$. As before we consider two cases.

If $\limsup_{r\to\infty} h(r) = \liminf_{r\to\infty} h(r)$, let m be the common value. If $m > 1$, let $\varepsilon < (m - 1)/2$. For all sufficiently large r, we have $G_{rr}/G \leqslant (1+\varepsilon)^2$. Now, from the differential equation for h, we have $h_r = G_{rr}/G - h^2 \leqslant (1+\varepsilon)^2 - (m-\varepsilon)^2 < 0$ for all sufficiently large r. Thus, $\limsup_{r\to\infty} h(r) = -\infty$, a contradiction, and thus $m \leqslant 1$.

In the second case we may assume that $\liminf_{r\to\infty} h(r) < \limsup_{r\to\infty} h(r)$. If the latter is strictly greater than 1, we may construct an infinite sequence $r_k \to \infty$ with $h(r_k) = A$, $h_r(r_k) \geqslant 0$ where $A > 1$. Choose $\varepsilon < A - 1$; for all sufficiently large r, we have $G_{rr}/G \leqslant (1+\varepsilon)^2$. Now, from the differential equation we have $h_r(r_k) = G_{rr}/G - h^2(r_k) \leqslant (1+\varepsilon)^2 - A^2 < 0$, a contradiction. Therefore, $\limsup_{r\to\infty} h(r) \leqslant 1$, which was to be proved.

REMARK 1. Lemmas (2.2)-(2.3) show in particular that if $\lim_{r\to\infty} K(r,\theta) = -k^2$, then $\lim_{r\to\infty} (G_r/G)(r,\theta) = |k|$. The proof shows further that if the first limit is uniform in $\theta \in [0,2\pi]$, then the second one is uniform also.

REMARK 2. The converse is not true in general. Indeed, suppose that G is of the form

$$G(r) = r \exp\left(\int_0^r p(s)\,ds\right),$$

where p is a C^1 function with $p(r) = 0$ for $0 \leqslant r \leqslant r_0$ and $p(r) = \sin^2(r^2)/4r$ for $r \geqslant r_1$. Then we have

$$G_r/G = 1 + \sin^2(r^2)/4r \qquad r > r_1$$

$$G_{rr}/G = 1 + \sin(r^2)\cos(r^2) + O(1/r) \qquad r \to \infty.$$

Then $\lim_{r\to\infty} G_r/G = 1$, but $\limsup_{r\to\infty} G_{rr}/G = 3/2$, $\liminf_{r\to\infty} G_{rr}/G = 1/2$.

REMARK 3. The proof of lemmas (2.2)-(2.3) shows that if $K(r,\theta)$ oscillates sufficiently slowly when $r \to \infty$, we may have $\liminf_{r\to\infty} \sqrt{-K} = \liminf_{r\to\infty} G_r/G < \limsup_{r\to\infty} G_r/G = \limsup_{r\to\infty} \sqrt{-K}$.

3. APPLICATIONS TO BROWNIAN MOTION.

Let $\{X_t, t \geq 0\}$ be the diffusion on M corresponding to the Riemannian metric (2.1). By a standard result [2], $X_t = (r(t),\theta(t))$ can be determined as the unique solution of the stochastic differential equations

$$r(t) = r(0) + \sqrt{2}\, w_1(t) + \int_0^t (G_r/G)(r(s),\theta(s))ds, \qquad (3.1a)$$

$$\theta(t) = \theta(0)+\sqrt{2}\int_0^t G^{-1}(r(s),\theta(s))dw_2(s)-\int_0^t (G_\theta/G^3)(r(s),\theta(s))ds, \qquad (3.1b)$$

where $(w_1(t),w_2(t))$ is a 2-dimensional Wiener process. We consider various conditions on the curvature at infinity.

$$\liminf_{r\to\infty} (-K) \geq k_-^2 > 0, \qquad (\text{uniformly},\ 0 \leq \theta \leq 2\pi). \qquad (H_1)$$

$$\limsup_{r\to\infty} (-K) \leq k_+^2 < \infty, \qquad (\text{uniformly},\ 0 \leq \theta \leq 2\pi). \qquad (H_2)$$

$$\lim_{r\to\infty} (-K) = k^2 > 0, \qquad (\text{uniformly},\ 0 \leq \theta \leq 2\pi). \qquad (H_3)$$

$$\lim_{r\to\infty} \frac{\int_0^r -KG dr}{\int_0^r G dr} = k^2 > 0, \qquad (\text{uniformly},\ 0 \leq \theta \leq 2\pi). \qquad (H_4)$$

We formulate the different a.s. asymptotic behaviors of $r(t)$, $t \to \infty$.

$$\liminf_{t\to\infty} r(t)/t \geq k_-. \tag{C_1}$$

$$\limsup_{t\to\infty} r(t)/t \leq k_+. \tag{C_2}$$

$$\lim_{t\to\infty} r(t)/t = k. \tag{C_3}$$

THEOREM 3.1. (H_1) *implies* (C_1); (H_1), (H_2) *imply* (C_2); *either* (H_3) *or* (H_4) *implies* (C_3).

PROOF. To prove the first statement, we must first show that $r(t) \to \infty$. To see this, note that Lemmas 2.1, 2.2 together imply that $G_r/G \geq \delta$ $(0 < r < \infty)$ for some positive constant δ. Thus, from (3.1a) we see that $r(t) \geq r(0) + w_1(t) + \delta t$. Hence, $r(t) \to \infty$. Now, return to (3.1a) and use (H_1) and Lemma 2.2 again on the integral term: $(G_r/G)(r(s),\theta(s)) \geq k_- - \varepsilon$ for $s \geq T$. Hence, $\liminf_{t\to\infty} r(t)/t \geq k_- - \varepsilon$ for any $\varepsilon > 0$.

To prove the second statement, apply Lemma 2.3 to the integral term in (3.1a). Since we already know that $r(t) \to \infty$, we see that $\limsup_{t\to\infty} r(t)/t \leq k_+$, as was to be proved.

To prove the final statement, note that (H_3) implies (H_4). Therefore we need only prove that (H_4) implies (C_3). To do this, define $H = G/\int_0^r G\,dr$. Then

$$H_r = G_r/\int_0^r G\,dr - H^2.$$

By the differential equation $G_{rr} + KG = 0$, we see that $\int_0^r KG\,dr = G_r(0,\theta) - G_r(r,\theta)$. Thus (H_4) states that $G_r/\int_0^r G\,dr \to k^2$. Therefore, $H_r = \overline{K} - H^2$, where $\lim_{r\to\infty} \overline{K} = k^2$.

Using the proof of lemmas (2.2)-(2.3), it follows that $\lim_{r\to\infty} H = k > 0$. Finally, we have $G_r/G = H^{-1}G_r/\int_0^r G dr \to k^{-1}k^2 = k$. Substituting this into (3.1a), we see that $r(t)/t \to k$.

REMARK 1. If we assume that the limits are uniform in θ and angle-dependent, i.e.

$$\liminf_{r\to\infty} (-K(r,\theta)) = k_-(\theta)^2, \qquad (H_1')$$

$$\limsup_{r\to\infty} (-K(r,\theta)) = k_+(\theta)^2, \qquad (H_2')$$

$$\lim_{r\to\infty} (-K(r,\theta)) = k^2(\theta) > 0, \qquad (H_3')$$

$$\lim_{r\to\infty} \frac{\int_0^r -K(r,\theta)G(r,\theta)d\theta}{\int_0^r G(r,\theta)d\theta} = k^2(\theta) > 0, \qquad (H_4')$$

then we may prove an analogue of Theorem 3.1. To do this we must first establish convergence of θ_t when $t \to \infty$. This was done by Prat [2] under the supplementary hypothesis $|G_\theta/G| \leq M$. In this case the conclusion (C_3) will read $\lim_{t\to\infty} r(t)/t = k^2(\theta_\infty)$, where $\theta_\infty = \lim_{t\to\infty} \theta_t$. For another proof, see [1, Section 3.1].

REMARK 2. Following the methods of [2-5], the preceeding results can be extended to dimension $n > 2$. The main result now takes the form $\lim_{t\to\infty} r(t)/t = (n - 1)k$, where k^2 is the common limit of the negative section curvatures when $r \to \infty$.

4. APPLICATION TO THE SPECTRUM OF THE LAPLACIAN.

Let λ_1 be the lower bound of the spectrum of the Laplace-Beltrami operator on M, i.e.

$$\lambda_1 = \inf_{f\neq 0} \frac{\int (f_r^2 + G^{-2}f_\theta^2)G\,dr\,d\theta}{\int f^2 G\,dr\,d\theta}, \qquad (4.1)$$

where f ranges over all C^∞ functions with compact support and the integration is over $[0,\infty) \times [0,2\pi]$. In [6], it was shown that $K \leqslant -k^2 < 0$ implies $\lambda_1 \geqslant k^2/4$. Let us use the preceeding estimates to obtain a qualitative result.

COROLLARY 4.1. *Under the hypothesis* (H_4), *we have* $\lambda_1 > 0$.

PROOF. The proof of Theorem (3.1) shows that $\lim_{r\to\infty} G_r/G = k > 0$. Together with Lemma (2.1), we have a global bound $G_r/G \geqslant \delta > 0$. Following [6], we write

$$\int_0^\infty f^2 G\,dr \leqslant \frac{1}{\delta}\int_0^\infty f^2 G_r\,dr$$

$$= -\frac{2}{\delta}\int_0^\infty f\,f_r G\,dr.$$

Using Schwarz's inequality, we have

$$\left(\int_0^\infty f^2 G\,dr\right)^2 \leqslant \frac{4}{\delta^2}\left(\int_0^\infty f^2 G\,dr\right)\left(\int_0^\infty f_r^2 G\,dr\right), \qquad (4.2)$$

$$\int_0^\infty f^2 G\,dr \leqslant \frac{4}{\delta^2}\int_0^\infty f_r^2 G\,dr. \qquad (4.3)$$

Adding the term $G^{-1}f_\theta^2$ to the right-hand member of (4.3) and integrating over $[0,2\pi]$, we see that the right-hand member of (4.1) is bounded below by $\delta^2/4$.

REMARK. Corollary 4.1 shows that the strict positivity of λ_1 can be inferred from the behavior of the curvature when $r \to \infty$. It is not true, however, that a lower bound for λ_1 can be found from the hypothesis (H_4). In another work [7], we have determined upper and lower bounds for λ_1 in terms of the ratios G_r/G, G_{rr}/G.

REFERENCES

1. P. Malliavin, *Diffusions et Géométrie Différentielle Globale*, Centro Internationale Mathematico, Estivo, August 1975.

2. J. J. Prat, *Etude asymptotique du mouvement brownien sur une variété riemannienne à courbure négative*, Comptes Rendus Academie des Sciences, Paris, 272 (1971), 1586-1589.

3. R. Azencott, *Behavior of Diffusions semigroups at infinity*, Bull. Soc. Math. France, 102 (1974), 193-240.

4. J. Vauthier, *Diffusions sur une variété riemannienne complete à courbure negative*, Comptes Rendus Academie des Sciences, Paris, 275 (1972), 925-926.

5. A. Debiard, B. Gaveau, E. Mazet, *Temps d'arrêt des diffusions riemanniennes*, Comptes Rendus Academie des Sciences, Paris, 278 (1974), 723-725.

6. H. P. McKean, *An upper bound to the spectrum of the Laplacian on a manifold of negative curvature*, J. Differential Geometry 4 (1970), 359-366.

7. M. Pinsky, *The spectrum of the Laplacian on a manifold of negative curvature*, (preprint).

Department of Mathematics
Northwestern University
Evanston, Illinois

Research supported by the National Science Foundation MPS71-02838-A04.

Two LIMIT THEOREMS FOR RANDOM EVOLUTIONS HAVING NON-ERGODIC DRIVING PROCESSES

BY

D. Stroock

In a forthcoming article [3], it is shown how the characterization of Markov processes in terms of the martingale problem (cf. [5]) can be used to prove limit theorems of the sort which arise in the study of random evolutions. In particular, one is able, in this way, to establish most of the beautiful theorems of Kurtz [2]. Moreover, what makes the martingale formulation valuable is that it enables one to handle some situations to which Kurtz's results are not applicable. Put briefly, what limit theorems for random evolutions have turned on is the fact that if one process $x(\cdot)$ is driven by a second process $y(\cdot)$ and if the process $y(\cdot)$ is sufficiently ergodic, then by speeding up both processes and making an appropriate scale change on $x(\cdot)$, one is able to make $x(\cdot)$ "see" $y(\cdot)$ as if $y(\cdot)$ had already achieved its equilibrium state. The net result is that in the limit, the process $x(\cdot)$ becomes autonomous (i.e. Markovian).

An architypical example of the preceding phenomenon is the following example, studied originally by Kac [1] and later by Pinsky [4]. Let $y(\cdot)$ be the Poisson process on $\{-1,1\}$ such that $P(y(0) = 1) = 1$ and $P(y(s + h) = y(s),$ $0 \leqslant h \leqslant t|\ y(u), \quad 0 \leqslant u \leqslant s) = e^{-t}$ and set $x(t) = \int_0^t y(s)ds$. Then for $\varepsilon > 0$, the process $z_\varepsilon(\cdot) = (\varepsilon x(\cdot), y(\cdot))$ is characterized by the facts that $P(z_\varepsilon(0) = (0,1)) = 1$ and for all test functions $\phi: \mathbb{R} \times \{-1,1\} \to \mathbb{R}$:

$$\phi(z_\varepsilon(t)) - \varepsilon\int_0^t y(s)\, \frac{\partial \phi}{\partial x}\,(z_\varepsilon(s))ds - \int_0^t K\phi(z_\varepsilon(s))ds$$

is a martingale, where $K\phi(x,y) = \phi(x,-y) - \phi(x,y)$. In particular, if $f \in C_0^\infty(\mathbb{R})$ and $\phi(x,y) = \frac{1}{2}\, yf'(x)$, then

$$\phi(z_\varepsilon(t)) - \frac{\varepsilon}{2} \int_0^t f''(\varepsilon x(s))ds + \int_0^t y(s)f'(\varepsilon x(s))ds$$

is a martingale, since $y^2 = 1$ and $K\phi(x,y) = -yf'(x)$. Multiplying through by ε, using $\varepsilon\int_0^t y(s)f'(\varepsilon x(s))ds$ $= f(\varepsilon x(t)) - f(0)$, and replacing t by t/ε^2, we conclude that:

$$f(\varepsilon x(t/\varepsilon^2)) - \frac{1}{2} \int_0^t f''(\varepsilon x(s/\varepsilon^2))ds + \varepsilon\phi(z_\varepsilon(t/\varepsilon^2))$$

is a martingale. If we now let P_ε on $C([0,\infty),\mathbb{R})$ be the distribution on $\varepsilon x(\cdot/\varepsilon^2)$, it is not hard to prove from this that $\{P_\varepsilon: \varepsilon > 0\}$ is pre-compact in the weak topology and that every limit point of $\{P_\varepsilon: \varepsilon > 0\}$ as $\varepsilon \searrow 0$ has the property that $x(0) = 0$ a.s. and $f(x(t)) - \frac{1}{2} \int_0^t f''(x(s))ds$ is a martingale for all $f \in C_0^\infty(\mathbb{R})$. Since these two properties characterize Wiener measure $\mathcal{W}$, we conclude that $P_\varepsilon \to \mathcal{W}$ weakly as $\varepsilon \searrow 0$.

In analyzing what makes the preceding work, one realizes that our ability to solve the equation $K\phi(x,y) = yf'(x)$ for any $f \in C_0^\infty(R)$ was essential, and it is here that the strong ergodicity of $y(t)$ enters (because range $(K) = \text{null}(\mu_0)$, where $\mu_0 = \frac{1}{2}(\delta_{\{-1\}} + \delta_{\{1\}})$ is the equilibrium distribution of $y(\cdot)$).

What happens in situations having a non-ergodic driving process $y(\cdot)$? Obviously, one can no longer expect the $y(\cdot)$ process to "wash out" and have the limiting $x(\cdot)$ process autonomous. This means that in order to describe the distribution of the limiting $x(\cdot)$ process, one has to keep track of the $y(\cdot)$ process as well. We will now investigate two examples of what can happen; other examples can be found in [3].

Let $\sigma: R^m \times R^n \to R^m \otimes R^n$ be a bounded smooth function having bounded first derivatives. Let $\beta(\cdot)$ be an n-dimensional Brownian motion, and for $\varepsilon > 0$ define $x_\varepsilon(\cdot)$ by

$$x_\varepsilon(t) = \varepsilon^{1/2}\int_0^t \sigma(x_\varepsilon(s),\beta(s))d\beta(s). \qquad (1)$$

Assume that there is a function $\bar{\sigma}(x,y)$ on $R^m \times R^n$ to $R^m \otimes R^n$ such that $\bar{\sigma}(x,y) = \bar{\sigma}\left(x, \frac{y}{|y|}\right)$, $|\bar{\sigma}(x_1,y) - \bar{\sigma}(x_2,y)| \leq A|x_1 - x_2|$, and $\int_{R^n} |\sigma(x,y) - \bar{\sigma}(x,y)|dy \leq B < \infty$. We will show that the distribution of $x_\varepsilon(\cdot/\varepsilon)$ tends weakly, as $\varepsilon \searrow 0$, to the distribution of the solution to

$$x(\cdot) = \int_0^\cdot \bar{\sigma}(x(s),\beta(s))d\beta(s). \qquad (2)$$

To see this, set $y_\varepsilon(\cdot) = \varepsilon^{1/2}\beta(\cdot)$ and $z_\varepsilon(\cdot) = (x_\varepsilon(\cdot),y_\varepsilon(\cdot))$. The distribution of $z_\varepsilon(\cdot)$ is uniquely characterized by the facts that $z_\varepsilon(0) = 0$ and for $\phi \in C_0^2(R^m \times R^n)$:

$$\phi(z_\varepsilon(t)) - \varepsilon\int_0^t L\phi(z_\varepsilon(s))ds \qquad (3)$$

is a martingale, where

$$L\phi(z) = \frac{1}{2}\sum_{i,j=1}^m (\sigma\sigma^*)_{ij}(x,y/\varepsilon^{1/2})\frac{\partial^2\phi}{\partial x_i\partial x_j} +$$

$$\sum_{i=1}^m\sum_{k=1}^n \sigma_{ik}(x,y/\varepsilon^{1/2})\frac{\partial^2\phi}{\partial x_i\partial y_k} + \frac{1}{2}\sum_{k=1}^n \frac{\partial^2\phi}{\partial y_k^2}. \qquad (4)$$

Given $f \in C_0^\infty(R^m \times R^n)$ and $\lambda > 0$, define

$$\phi(z) = \phi_f^\lambda(x,y) = \sum_{i,j=1}^m \psi_{ij}^\lambda(x,y/\varepsilon^{1/2})f_{,ij}(x,y)$$

$$+ \sum_{i=1}^m\sum_{k=1}^n \psi_{ik}^\lambda(x,y/\varepsilon^{1/2})f_{,ik}(x,y),$$

where $f_{,ij} = \frac{\partial^2 f}{\partial x_i\partial x_j}$, $f_{,ik} = \frac{\partial^2 f}{\partial x_i\partial y_k}$, and

$$\psi_{ij}^\lambda(x,y) = \frac{1}{2}[r_\lambda * ((\sigma\sigma^*)_{ij}(x,\cdot) - (\bar{\sigma}\,\bar{\sigma}^*)_{ij}(x,\cdot))](y),$$

$$\psi_{ik}^\lambda(x,y) = [r_\lambda * (\sigma_{ik}(x,\cdot) - \bar{\sigma}_{ik}(x,\cdot))](y),$$

with $r_\lambda(y) \equiv \int_0^\infty \frac{1}{(2\pi t)^{n/2}} e^{-\lambda t - |y|^2/2t}dt$. In order to eliminate unnecessary mess, we will show only in the case $m = n = 1$ how this choice of test function yields the desired result (the general case is more tedious, but no more difficult). Note that:

$$L\phi_f^\lambda(z) = -\frac{1}{\varepsilon}\left(\frac{1}{2}\sigma^2(x,y/\varepsilon^{1/2})f_{,11}(z) + \sigma(x,y/\varepsilon^{1/2})f_{,12}(z)\right)$$

$$+ \frac{1}{\varepsilon}\left(\frac{1}{2}\overline{\sigma}^2(z)f_{,11}(z) + \overline{\sigma}(z)f_{,12}(z)\right)$$

$$+ I_\varepsilon^\lambda(z) + \frac{1}{\varepsilon^{1/2}} J_\varepsilon^\lambda(z) + \frac{1}{\varepsilon} K_\varepsilon^\lambda(z),$$

where

$$I_\varepsilon^\lambda(z) = \frac{1}{2}\sigma^2(x,y/\varepsilon^{1/2})\phi_{,11}(z) + \frac{1}{2}[\psi_{11}^\lambda(x,y/\varepsilon^{1/2})f_{,1122}(z) + \psi_{12}^\lambda(x,y/\varepsilon^{1/2})f_{,1222}(z)],$$

$$J_\varepsilon^\lambda(z) = \sigma(x,y/\varepsilon^{1/2})\phi_{,12}(z) + [\psi_{11,2}^\lambda(x,y/\varepsilon^{1/2})f_{,112}(z) + \psi_{12,2}^\lambda(x,y/\varepsilon^{1/2})f_{,122}(z)],$$

$$K_\varepsilon^\lambda(z) = \lambda[\psi_{11}^\lambda(x,y/\varepsilon^{1/2})f_{,11}(z) + \psi_{12}^\lambda(x,y/\varepsilon^{1/2})f_{,12}(z)].$$

Thus,

$$\varepsilon\phi_f^\lambda(z_\varepsilon(t)) + \varepsilon\int_0^t Lf(z_\varepsilon(s))ds - \varepsilon\int_0^t \overline{L}f(z_\varepsilon(s))ds$$

$$- \varepsilon^2\int_0^t I_\varepsilon^\lambda(z_\varepsilon(s))ds - \varepsilon^{3/2}\int_0^t J_\varepsilon^\lambda(z_\varepsilon(s))ds$$

$$- \varepsilon\int_0^t K_\varepsilon^\lambda(z_\varepsilon(s))ds$$

is a martingale, where

$$\overline{L} = \frac{1}{2}\,\overline{\sigma}^2(z)\,\frac{\partial^2}{\partial x^2} + \overline{\sigma}(z)\,\frac{\partial^2}{\partial x \partial y} + \frac{1}{2}\,\frac{\partial^2}{\partial y^2}.$$

Also,

$$f(z_\varepsilon(t)) - \varepsilon \int_0^t Lf(z_\varepsilon(s))ds$$

is a martingale. Hence,

$$f(z_\varepsilon(t)) - \varepsilon \int_0^t \overline{L}f(z_\varepsilon(s))ds - \varepsilon^2 \int_0^t I_\varepsilon^\lambda(z_\varepsilon(s))ds$$

$$- \varepsilon^{3/2} \int_0^t J_\varepsilon^\lambda(z_\varepsilon(s))ds - \varepsilon \int_0^t K_\varepsilon^\lambda(z_\varepsilon(s))ds$$

is a martingale, and therefore so is

$$f(z_\varepsilon(t/\varepsilon)) - \int_0^t \overline{L}f(z_\varepsilon(s/\varepsilon))ds - \varepsilon \int_0^t I_\varepsilon^\lambda(z_\varepsilon(s/\varepsilon))ds$$

$$- \varepsilon^{1/2} \int_0^t J_\varepsilon^\lambda(z_\varepsilon(s/\varepsilon))ds - \int_0^t K_\varepsilon^\lambda(z_\varepsilon(s/\varepsilon))ds.$$

Note that for fixed λ, $\varepsilon I_\varepsilon^\lambda$ and $\varepsilon^{1/2} J_\varepsilon^\lambda \to 0$ uniformly in z as $\varepsilon \searrow 0$. Also, $K_\varepsilon^\lambda \to 0$ uniformly in z and $\varepsilon > 0$ as $\lambda \searrow 0$. Hence if P_ε is the distribution of $z_\varepsilon(\cdot/\varepsilon)$, then $\{P_\varepsilon : \varepsilon > 0\}$ is pre-compact in the weak topology and every limit P as $\varepsilon \searrow 0$ of P_ε has the properties that $P(z(0) = 0) = 1$ and

$$f(z(t)) - \int_0^t \overline{L}f(z(s))ds$$

is a martingale for all $f \in C_0^\infty(R^2)$. It is easily seen from this that P_ε tends weakly as $\varepsilon \searrow 0$ to the distribution of

$(x(\cdot),\beta(\cdot))$, where $\beta(\cdot)$ is a Brownian motion and $x(\cdot) = \int_0^{\cdot}\overline{\sigma}(x(s),\beta(s))d\beta(s)$.

Before dropping this example, one should observe that the same technique can be used to prove the following result. Let $Y_1,\cdots,Y_k,\cdots$ be i.i.d. random variable having mean 0 and variance 1. Given a smooth function $\sigma\colon R^1 \to R^1$ with the property that $\int_{R^1}|\sigma(y) - \overline{\sigma}(y)|dy < \infty$ for some function $\overline{\sigma}(\cdot)$ satisfying $\overline{\sigma}(y) = \overline{\sigma}(y/|y|)$, set

$$X_n = \sum_{m=1}^{n} \sigma\left(\sum_{k=1}^{m-1} Y_k\right) Y_m$$

and let $x_\varepsilon(t) = \varepsilon^{1/2}X_{[nt/\varepsilon]}$. Then the distribution of $x_\varepsilon(\cdot)$ tends to the distribution of the process $x(\cdot)$ in (2).

Next, consider the process $x_\varepsilon(\cdot)$ given by:

$$x_\varepsilon(\cdot) = \varepsilon^{1/2}\int_0^{\cdot}\sigma(x_\varepsilon(s),\beta(s))d\beta(s),$$

where $\beta(\cdot)$ is a 1-dimensional Brownian motion and $\sigma\colon R^m \times R^1 \to R^m \otimes R^1$ is a bounded smooth function satisfying $\sigma(\cdot,y) \equiv 0$ for large $|y|$. By the preceding, we know that $x_\varepsilon(\cdot/\varepsilon)$ tends in probability, uniformly on finite time intervals, to 0 as $\varepsilon \searrow 0$. What we are going to show now is that if $a(x) \equiv \int_{-\infty}^{\infty}\sigma\sigma^*(x,y)dy$, $x(\cdot)$ is determined by

$$x(\cdot) = \int_0^{\cdot}a^{1/2}(x(s))d\tilde{\beta}(s),$$

where $\tilde{\beta}(\cdot)$ is an m-dimensional Brownian motion, and $\ell_0(\cdot)$ is the local time at 0 of a 1-dimensional Brownian motion $\gamma(\cdot)$ which is independent of $\tilde{\beta}(\cdot)$, then the distribution of $x_\varepsilon(\cdot/\varepsilon^2)$ tends weakly to that of $x(\ell_0(\cdot))$. In order to do this, first observe that the distribution of the pair $(x(\ell_0(\cdot)),\gamma(\cdot))$ just described is characterized by the facts that $(x(0), \gamma(0)) = (0,0)$ and

$$f(x(\ell_0(t)),\gamma(t))-\int_0^t \bar{L}f(x(s),\gamma(s))d\ell_0(s)- \frac{1}{2}\int_0^t \frac{\partial^2 f}{\partial y^2}(x(s),\gamma(s))ds \quad (5)$$

is a martingale for all $f \in C_0^\infty(R^m \times R^1)$, where

$$\bar{L} = \frac{1}{2}\sum_{i,j=1}^m a_{ij}(x) \frac{\partial^2}{\partial x_i \partial x_j},$$

and $\ell_0(\cdot)$ is the local time of $\gamma(\cdot)$ at 0. Next note that the distribution of $z_\varepsilon(\cdot) \equiv (x_\varepsilon(\cdot),\varepsilon\beta(\cdot))$ is characterized by $z_\varepsilon(0) = (0,0)$ and

$$\phi(z_\varepsilon(t)) -\varepsilon\int_0^t L\phi(z_\varepsilon(s))ds \quad (6)$$

is a martingale for all $\phi \in C_k^2(R^m \times R^1)$, where

$$L\phi = \frac{1}{2}\sum_{i,j=1}^m (\sigma\sigma^*)_{ij}(x,y/\varepsilon) \frac{\partial^2\phi}{\partial x_i \partial x_j}$$

$$+ \varepsilon^{1/2}\sum_{i=1}^m \sigma_i(x,y/\sigma) \frac{\partial^2\phi}{\partial x_i \partial y} + \varepsilon \frac{\partial^2\phi}{\partial y^2}.$$

Define $\psi_{ij}(z) = -\int_{-\infty}^y d\eta \int_{-\infty}^\eta [(\sigma\sigma^*)_{ij}(x,\xi) - \rho(\xi)a_{ij}(x)]d\xi$, $1 \leqslant i,j \leqslant m$, where $\rho \in C_0^\infty(-1,1)$ is a non-negative function such that $\int_{-\infty}^\infty \rho(y)dy = 1$. For $f \in C_0^\infty(R^m \times R^1)$ set

$$\phi(z) = \phi_f(x,y) = \sum_{i,j=1}^m \psi_{ij}(x,y/\varepsilon) \frac{\partial^2 f}{\partial x_i \partial x_j}(z).$$

For convenience, we again will carry out the rest of the proof only for the case $m = 1$. Applying (6) to $\phi_f(\cdot)$, we get that

$$\begin{aligned}
\phi(z_\varepsilon(t)) &- \frac{\varepsilon}{2}\int_0^t \sigma^2(x_\varepsilon(s),\beta(s))\phi_{,11}(z_\varepsilon(s))ds \\
&- \varepsilon^{3/2}\int_0^t \sigma(x_\varepsilon(s),\beta(s))\phi_{,12}(z_\varepsilon(s))ds \\
&- \frac{\varepsilon^2}{2}\int_0^t \psi(x_\varepsilon(s),\beta(s))f_{,1122}(z_\varepsilon(s))ds \\
&- \varepsilon\int_0^t \psi_{,2}(x_\varepsilon(s),\beta(s))f_{,112}(z_\varepsilon(s))ds \\
&+ \frac{1}{2}\int_0^t \sigma^2(x_\varepsilon(s),\beta(s))f_{,11}(z_\varepsilon(s))ds \\
&- \frac{1}{2}\int_0^t a(x_\varepsilon(s))f_{,11}(z_\varepsilon(s))\rho(\beta(s))ds
\end{aligned}$$

is a martingale. Also,

$$\begin{aligned}
f(z_\varepsilon(t)) &- \frac{\varepsilon}{2}\int_0^t \sigma^2(x_\varepsilon(s),\beta(s))f_{,11}(z_\varepsilon(s))ds \\
&- \varepsilon^{3/2}\int_0^t \sigma(x_\varepsilon(s),\beta(s))f_{,12}(z_\varepsilon(s))ds \\
&- \frac{\varepsilon^2}{2}\int_0^t f_{,22}(z_\varepsilon(s))ds
\end{aligned}$$

is a martingale. Thus, if $\rho_\varepsilon(y) = \frac{1}{\varepsilon}\rho(y/\varepsilon)$, then

$$\begin{aligned}
f(z_\varepsilon(t/\varepsilon^2)) &- \frac{1}{2}\int_0^t a(x_\varepsilon(s/\varepsilon^2))f_{,11}(z_\varepsilon(s/\varepsilon^2))\rho_\varepsilon(y_\varepsilon(s/\varepsilon^2))ds \\
&- \frac{1}{2}\int_0^t f_{,22}(z_\varepsilon(s/\varepsilon^2))ds \\
&- \frac{1}{2}\int_0^t \sigma^2(x_\varepsilon(s/\varepsilon^2),\beta(s/\varepsilon^2))\phi_{,12}(z_\varepsilon(s/\varepsilon^2))ds \\
&- \int_0^t \psi_{,2}(x_\varepsilon(s/\varepsilon^2),\beta(s/\varepsilon^2))f_{,112}(z_\varepsilon(s/\varepsilon^2))ds
\end{aligned}$$

$$- \varepsilon^{1/2} \int_0^t \sigma(x_\varepsilon(s/\varepsilon^2), \beta(s/\varepsilon^2))\phi_{,12}(z_\varepsilon(s/\varepsilon^2))ds$$

$$- \varepsilon \int_0^t \psi(x_\varepsilon(s/\varepsilon^2), \beta(s/\varepsilon^2))f_{,1122}(z_\varepsilon(s))ds$$

$$+ \varepsilon\phi(z_\varepsilon(t/\varepsilon^2))$$

is a martingale. It is easy to see that all the terms, other than the first three, tend in the mean to zero as $\varepsilon \searrow 0$. Also, one can show that if P_ε is the distribution of $z_\varepsilon(\cdot/\varepsilon^2)$, then $\{P_\varepsilon: \varepsilon > 0\}$ is pre-compact in the weak topology. Now let $\varepsilon_n \searrow 0$ be a sequence for which $P_n = P_{\varepsilon_n}$ converges weakly to P as $n \to \infty$. Note that $y(\cdot)$ is a 1-dimensional Brownian motion under P_n for all n, and therefore it is also a 1-dimensional Brownian motion under P. Let $\ell_0(\cdot)$ be the local time of $y(\cdot)$ at 0. Then what we must show is that

$$f(z(t)) - \frac{1}{2} \int_0^t a(x(s))f_{,11}(z(s))d\ell_0(s) - \frac{1}{2} \int_0^t f_{,22}(z(s))ds$$

is a martingale under P for all $f \in C_0^\infty(R^2)$. It is easy to see that all one needs in order to show this is that

$$E^{P_n}[\Phi\int_0^t a(x(s))f_{,11}(z(s))\rho_{\varepsilon_n}(y(s))ds]$$

$$\to E^P[\Phi\int_0^t a(x(s))f_{,11}(z(s))\ell_0(ds)]$$

for all bounded continuous Φ on $C([0,\infty),R^2)$. But for any $\delta > 0$

$$\overline{\lim_{n\to\infty}} |E^{P_n}[\Phi \int_0^t a(x(s))f_{,11}(z(s))\rho_{\varepsilon_n}(y(s))ds] - E^P[\Phi \int_0^t a(x(s))f_{,11}(z(s))\ell_0(ds)]|$$

$$\leq \overline{\lim_{n\to\infty}} |E^{P_\varepsilon}[\Phi \int_0^t a(x(s))f_{,11}(z(s))(\rho_{\varepsilon_n}(y(s)) - \rho_\delta(y(s)))ds]| + |E^P[\Phi \int_0^t a(x(s))f_{,11}(z(s))\rho_\delta(y(s))ds] - E^P[\Phi \int_0^t a(x(s))f_{,11}(z(s))\ell_0(ds)]|,$$

and the second term on the right tends to 0 as $\delta \searrow 0$. Thus, it suffices to show that if $\phi \in C_0(\mathcal{R}^2)$, then

$$\lim_{\delta\searrow 0} \sup_{\varepsilon>0} E^{P_\varepsilon}[|\int_0^t \phi(z(s))\rho_\delta(y(s))ds - \int_0^t \phi(z(s))\ell_0(ds)|] = 0.$$

Note that $\int_0^t \phi(z(s))\rho_\delta(y(s))ds = \int \rho_\delta(y)(\int_0^t \phi(x(s),y)\ell_y(ds))dy$, where $\ell_y(\cdot)$ is the local time at y of $y(\cdot)$. Thus, it is enough to show that for $\phi \in C_0(\mathcal{R})$

$$\lim_{\delta\searrow 0} \sup_{\varepsilon>0} E^{P_\varepsilon}[|\int \rho_\delta(y)(\int_0^t \phi(x(s))(\ell_y(ds) - \ell_0(ds)))dy|] = 0.$$

Since $\{P_\varepsilon: \varepsilon > 0\}$ is tight, it is easy to see that for each $\alpha > 0$, there is a compact $K \subseteq C([0,\infty),\mathcal{R}^2)$ such that

$$E^{P_\varepsilon}[|\int \rho_\delta(y)(\int_0^t \phi(x(s))(\ell_y(ds) - \ell_0(ds)))dy|, K^c] < \alpha$$

for all $\delta > 0$ and $\varepsilon > 0$. Observe that there is a bounded, continuous, decreasing function μ_K on $[0,\infty)$ such that $\mu_K(0) = 0$ and for all $\omega \in K$ and all continuous functions $v: [0,t] \to \mathcal{R}$ having bounded variation,

$$|\int_0^t \phi(x(s,\omega))v(ds)| \leq \mu_K(\sup_{0\leq s\leq t} |v(s)|)(1 + |v|(t)),$$

where $|v|(t)$ is the total variation of v on $[0,t]$. Thus, if $v_\delta(\cdot) = \int \rho_\delta(y)(\ell_y(\cdot) - \ell_0(\cdot))dy$, then for all $\varepsilon > 0$ and $\delta > 0$

$$E^{P_\varepsilon}[|\int \rho_\delta(y)(\int_0^t \phi(x(s))(\ell_y(ds) - \ell_0(ds)))dy)]$$

$$\leq \alpha + E^{P_\varepsilon}[(1 + |v_\delta|(t))\mu_K(\sup_{0\leq s\leq t} |v_\delta(s)|)].$$

Since the distribution of $v_\delta(\cdot)$ under P_ε is independent of $\varepsilon > 0$ and because $\ell_y(s)$ is jointly continuous with respect to (s,y), it is easy to see that

$$\lim_{\delta\searrow 0} \sup_{\varepsilon>0} E^{P_\varepsilon}[(1 + |v_\delta|(t))\mu_K(\sup_{0\leq s\leq t} |v_\delta(s)|)] = 0.$$

REFERENCES

1. M. Kac, *Some Stochastic Problems in Physics and Mathematics*, Magnolia Petrolium Lectures in Pure and Applied Science, 2 (1956).

2. T. Kurtz, *A Limit Theorem for Perturbed Operator Semi-groups with Applications to Random Evolutions*, J. Fnal. Anal. 12 (1973).

3. G. Papanicolaou, D. Stroock and S. R. S. Varadhan, forthcoming.

4. M. Pinsky, *Differential Equations with a Small Parameter and the Central Limit Theorem for Functions Defined on a Finite Markov Chain*, Z. Wahr. 9 (1968), 101-111.

5. D. Stroock and S. R. S. Varadhan, *Diffusions with Continuous Coefficients, I*, Comm. Pure Appl. Math. XXII (1969), 345-400.

Department of Mathematics
University of Colorado
Boulder, Colorado

N.S.F. # MP574-18925.

A
B 7
C 8
D 9
E 0
F 1
G 2
H 3
I 4
J 5

